Prolonged Fasting

A Systems Guide to Duration, Recovery, and Metabolic Limits

Nathan Reed

PERENNIAL PUBLISHING

Published by Perennial Publishing

Paperback ISBN: 978-1-971996-21-9
Hardcover ISBN: 978-1-971996-22-6
E-book ISBN: 978-1-971996-23-3

This book is intended for educational and informational purposes only. It does not constitute medical, nutritional, or health advice. The author and publisher make no representations or warranties regarding the completeness or accuracy of the information contained herein. Readers are responsible for their own health decisions and are encouraged to consult qualified healthcare professionals before making changes to diet, fasting practices, supplementation, exercise, or lifestyle. The author and publisher shall not be held liable for any loss, injury, or damages arising from the use of this material.

Contents

1. Duration as a System Variable 1
2. Event vs. Pattern 9
3. Magnitude 17
4. Thresholds 25
5. The Depletion Curve 35
6. Adaptive Shift 43
7. Downregulation 49
8. Cumulative Load 57
9. Hormonal Strain 65
10. Resource Conservation 73
11. Refeeding Sensitivity 81
12. Recovery Debt 87
13. Rebound Instability 93
14. Capacity Constraints 99
15. Aging Constraints 107
16. Error Margins & Compensation 115
17. Failure Cascades 123

18. Oscillation Envelope (CORE) 131

19. Envelope Matching 141

20. Frequency vs. Magnitude Tradeoff 149

21. Strategic Use 157

22. Strategic Non-Use 165

23. Magnitude Without Structure 173

Duration as a System Variable

Fasting is often grouped into a single category as though the absence of intake were the primary variable and all forms of fasting therefore differed only by degree. That framing collapses structurally distinct phenomena into a single idea. The more useful distinction is not whether intake is absent, but how absence is organized in time. Intermittent fasting is best understood as a pattern variable. Its defining feature is recurrence: feeding and non-feeding alternate in repeated cycles, allowing the organism to move through familiar transitions while preserving a broader rhythm of input, repletion, signaling, and recovery. Prolonged fasting is not this pattern carried further. It is governed instead by duration. Once duration becomes the dominant variable, the governing logic changes. The system is no longer primarily organizing around repetition and return, but around extended continuity of absence. That shift matters because biological systems do not respond only to condition. They respond to condition over time. A state that is tolerable, adaptive, or metabolically coherent over one interval may become qualitatively different when sustained beyond the range in which its assumptions still hold.

This distinction is foundational because repeating a shorter cycle and extending a single uninterrupted duration are not interchangeable operations. Repetition preserves alternation. Extension removes it. Alternation allows partial restoration of energy availability, endocrine signaling, substrate replenishment, and downstream repair processes that depend on periodic input. Duration extension does not simply deepen the same state while keeping its structure intact. It progressively changes the structure itself. What appears as "more fasting" is in fact a different temporal architecture with different constraints. The same absence of intake may be present in both cases, but its system meaning diverges once the interval of absence becomes long enough to alter the conditions under which regulation is being maintained. The question is therefore not whether the system is fasting, but what duration is doing to the system while it is fasting.

That is the central principle required to understand prolonged fasting clearly: system stress does not increase linearly with fasting duration. Duration escalation is nonlinear. This means that each extension in time does not simply add more of the same metabolic demand. Instead, additional time can shift the organism into different regulatory conditions, with different priorities, different compensations, and different costs. The error in treating fasting duration as a straight line is that it assumes continuity where biological systems often display threshold behavior, curvature, and phase change. A system can absorb one interval of deprivation within a relatively stable operating range and then begin to reorganize under a longer interval as internal reserves, signaling dynamics, and recovery assumptions are progressively altered. The absence of intake may remain constant, but the internal system state does not.

This is why later stages of fasting cannot be understood as mere extensions of earlier stages. Early fasting is not simply the first segment of prolonged fasting in a way that preserves equivalence. Early stages may still operate within a broader field of reversible and expected metabolic adjustment, where the organism is redistributing substrate use, modifying hormone exposure, and managing short-term absence without yet confronting deeper structural consequences. As duration extends, the organism is no longer only adjusting fuel preference or managing an interrupted input stream. It begins operating under a different burden of continuity. The question shifts from whether energy can be redistributed to whether redistribution can continue without compounding distortion. The relevant issue becomes not whether the system can remain active in the absence of intake, but what kind of activity it must increasingly perform to preserve function under prolonged absence. Duration changes the problem the system is solving.

This is the point at which stress must be understood as state-dependent rather than simply cumulative in a linear sense. Nonlinearity does not mean change appears suddenly. It means that the meaning of each additional increment of time depends on the state already produced by the preceding increments. A system that has already shifted substrate allocation, altered hormonal ratios, reduced incoming substrate diversity, and begun compensatory conservation is not equivalent to the same system at the outset of fasting. Each additional unit of duration is imposed on a system that has already been changed by the duration that came before it. The same extra interval therefore does not land on the same baseline. It lands on a moving structure. Stress rises not simply because more time has passed, but because time progressively changes the conditions under which further time must be endured.

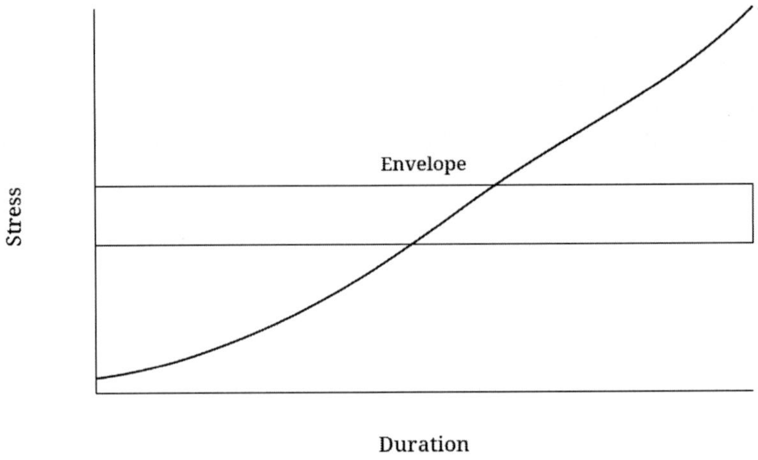

[Figure 1: Duration vs. System – Stress with Oscillation Envelope]

Once duration is recognized as nonlinear, prolonged fasting becomes less intelligible as a behavioral category and more intelligible as a stress architecture. The system is not simply continuing without food. It is maintaining coherence under an expanding field of constraints. These constraints do not emerge all at once, and they do not belong to a single domain. They accumulate in layers. Energy constraint is the most obvious layer because the absence of intake removes exogenous substrate and forces increasing reliance on internal reserves and reallocation. But energy alone does not explain the system burden of prolonged fasting. Biological regulation is not powered only by calories. It is organized through signaling, timing, allocation, suppression, compensation, and recovery. As duration extends, the organism must not only keep supplying tissues with usable fuel. It must also maintain regulatory coordination while doing so under altered conditions.

Signaling constraint emerges because prolonged absence of intake changes the informational environment of the organism. Input is not only material; it is regulatory. Feeding carries timing information, endocrine consequences, substrate-specific signaling, and rhythmic reinforcement for a wide array of downstream processes. When input remains absent for longer intervals, the system is not simply deprived of incoming energy. It is also operating with a different pattern of cues. That matters because stability depends not only on having enough substrate, but on having coherent signals about what state the organism is in and how long that state is expected to persist. Extended duration therefore introduces an informational strain. The system must infer continuity, reprioritize allocation, and preserve internal coordination while a major class of input-linked signals remains suppressed or absent.

Recovery constraint emerges because all systems that tolerate stress do so within assumptions about eventual restoration. A stressor is not defined only by what it demands in the moment. It is also defined by what must later be repaired, replenished, or re-stabilized after that demand has been carried. Prolonged fasting changes the recovery equation because duration determines how many domains are being borrowed against while intake remains absent. A shorter absence may be managed within a cycle that assumes timely re-entry into a fed condition. A longer absence extends the period during which the organism must continue to regulate without that return. The result is not simply a longer challenge, but a different relationship between demand and restoration. Recovery is no longer a background assumption. It becomes part of the structural cost of the event itself.

This is why the absence of input should never be mistaken for the absence of activity. Biological quietness is often inferred where visible intake has stopped, as though fasting represented a simplification of system behavior. In reality, prolonged fasting is metabolically active precisely because the organism must keep functioning while normal conditions of supply have been suspended. Fuel must still be mobilized, tissue demands still prioritized, signaling hierarchies still maintained, and protective compensations still coordinated. The system is not doing less because intake has ceased. In many respects, it is doing more with fewer degrees of freedom. The apparent simplicity of not eating conceals a growing internal workload. What disappears externally is input. What expands internally is regulatory burden.

As that burden extends across time, the organism must manage not only immediate survival but continuity of oscillation. Biological systems remain stable not by holding a single static position, but by moving within bounded ranges of variation. Hormones rise and fall, substrates shift, appetite signals fluctuate, autonomic tone changes, and recovery processes alternate with expenditure. Stability is therefore not stillness. It is organized oscillation within tolerable limits. Prolonged fasting becomes structurally significant because duration can begin to compress, distort, or overextend those oscillations. What matters is not only whether the system can remain in a fasting condition, but whether the amplitude, timing, and reversibility of its internal movements remain intact while doing so.

That boundary condition can be described as an oscillation envelope. It is not a single threshold and not a binary line between safety and failure. It is better understood as the dynamic range within which stress and recovery can continue to interact without producing escalating distortion. A system remains viable not merely because it can continue under stress, but because the stress remains containable within an envelope that still permits eventual re-stabilization. Once duration pushes the system toward the edge of that envelope, the problem is no longer simply one of persistence. It becomes one of narrowing adaptability. The organism may still function, but with reduced flexibility, greater compensatory strain, and a smaller margin for maintaining coherence across domains.

This model matters because it replaces simplistic duration thinking with a more structurally accurate one. The relevant question is not how long absence can be extended in abstract terms, nor whether a longer interval necessarily represents a deeper or more meaningful expression of fasting. Those framings confuse extension with value and duration with progression. A systems view does not treat longer exposure as inherently more productive simply because it is longer. It asks instead what new burdens duration introduces, how those burdens interact, and whether the resulting system behavior remains inside a recoverable range. Prolonged fasting therefore cannot be understood as a scaled-up version of intermittent fasting without losing the very variable that governs its consequences. Once duration becomes primary, the system must be read differently.

The rest of the book depends on that shift in framing. Prolonged fasting is not best interpreted as a cleaner, stricter, or more advanced form of abstention. It is a duration-dependent stress event whose structure changes as time extends. Its meaning lies not in the absence of intake, but in the nonlinear reorganization that duration imposes on energy handling, signaling coherence, compensatory behavior, and recovery capacity. To understand it clearly requires abandoning the assumption that fasting remains the same phenomenon as it lengthens. It does not. It becomes something more constrained, more state-dependent, and more structurally demanding. The oscillation envelope will serve as the governing model for interpreting that progression, because it captures the central reality that prolonged fasting is not defined only by what is removed, but by what the system must continue to preserve while removal persists.

Event vs. Pattern

The clearest way to distinguish intermittent fasting from prolonged fasting is not by intent, appearance, or severity, but by temporal structure. Biological systems do not only respond to what is absent or present. They respond to how conditions are organized across time. This makes rhythm a defining variable. Intermittent fasting is best understood as a pattern-based structure because its coherence depends on repetition. It is not simply a state of not eating that happens to recur. It is a recurring alternation between intake and non-intake, and that alternation is what gives it system meaning. Without recurrence, there is no pattern. Without return, there is no rhythm. The system does not interpret a repeating short-cycle absence the same way it interprets a single extended absence, even if both are superficially grouped under the same label of fasting.

A pattern is not just a sequence. It is a repeated temporal arrangement that allows the organism to anticipate, distribute, and partially restore across cycles. In a pattern-based fasting structure, the fasting period is metabolically meaningful only because it is embedded within repeated returns to feeding. Those returns matter because they interrupt continuity of absence and preserve the possibility of oscillation. Energy availability is therefore not defined only by whether energy is present at any given moment, but by whether the system can repeatedly move back into states of replenishment before accumulated demands become structurally dominant. This is one of the defining features of intermittent fasting as a pattern rather than an isolated condition. The organism is not simply deprived and then restored in a one-time event. It is moving through a repeatable loop in which partial restoration is built into the architecture itself.

That built-in return changes how the system distributes strain. A repeatable fasting pattern allows non-feeding to occur within a larger field of recurrence where substrate availability, hormonal signaling, and repair assumptions remain cyclically reinforced. The body is not required to remain indefinitely coherent under uninterrupted absence. It is allowed to alternate. That alternation does not remove all stress, nor does it guarantee stability in every context, but it changes the character of the stress by embedding it inside a rhythm of expected restoration. This matters because biological systems regulate more effectively when variation is bounded by repetition. Oscillatory systems tolerate movement precisely because movement is expected to reverse, recur, and remain coupled to broader cycles of input and output. Pattern, in this sense, is not merely repetition for its own sake. It is the structure that keeps stress from becoming continuous.

This is why intermittent fasting should be defined less by the fasting interval itself than by the recurrence of return. Return is not a secondary feature. It is part of the mechanism. A fasting pattern remains a pattern only if the organism repeatedly re-enters conditions that allow restoration of energy handling, re-engagement of feeding-linked signals, and reconstitution of the broader metabolic and endocrine environment in which the next cycle will occur. Recovery processes do not need to be completed in full after every cycle for a pattern to exist, but they must remain periodically available within the structure. The system is therefore not asked to sustain an unbroken deprivation event. It is asked to oscillate through repeated, partially restorative states. That oscillation is what makes pattern-based fasting structurally distinct from prolonged fasting.

Once that distinction is clear, prolonged fasting becomes easier to define accurately. It is not a pattern. It is an event. The defining feature of an event is not simply that it happens once, but that it does not derive its meaning from repeatable alternation. A prolonged fast is governed by duration rather than recurrence. Its structure is not built around cycles of absence and return, but around continuity of absence sustained across time. That continuity changes the governing logic of the system. When return is removed from the immediate architecture, the organism is no longer operating inside a bounded rhythm. It is operating inside a discrete stress event whose progression depends on how long the absence continues and what that duration begins to impose across multiple layers of regulation.

This distinction is essential because a prolonged fast is often misinterpreted as if it were merely an enlarged fasting interval within the same pattern logic. That interpretation fails because extending duration is not equivalent to repeating a cycle. Repetition preserves alternation. Duration extension removes it. In a pattern, the system repeatedly re-enters a fed state before continuity of absence becomes the primary organizing condition. In an event, the absence is not interrupted by built-in restoration. This changes not only the magnitude of strain, but the type of strain. The system is no longer modulating around a recurring loop. It is carrying forward a single uninterrupted condition whose consequences are cumulative, state-dependent, and increasingly constrained by time. What begins as absence becomes continuity, and continuity has different structural implications than repetition.

A prolonged fast therefore functions as a discrete stress event because it asks the organism to maintain internal coordination under sustained withdrawal of input without the restoring alternation that defines a pattern. Duration replaces recurrence as the governing variable. That replacement matters because time does not merely lengthen exposure. It changes the system operating environment. A condition that can be regulated within a recurring cycle may become structurally different when extended continuously. The organism must begin to solve new problems as duration increases, not simply more of the same problem. It must continue allocating fuel, preserving essential function, managing signaling coherence, and containing the downstream consequences of uninterrupted absence without the periodic return that would ordinarily help distribute and dissipate stress. This is what makes prolonged fasting event-based rather than pattern-based. Its meaning lies in its continuity.

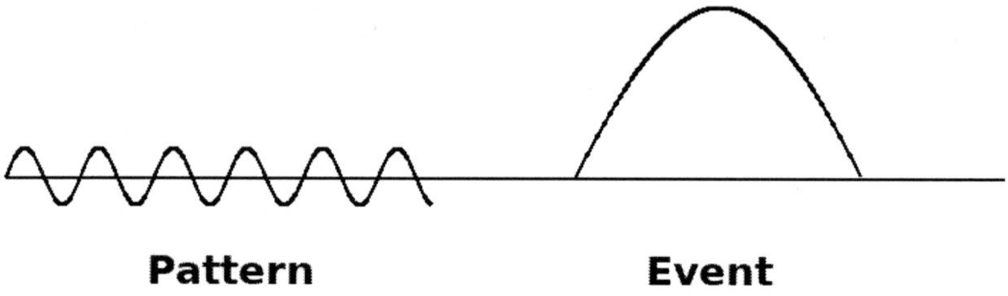

[Figure 2: Pattern waves vs single spike event—repeating oscillations contrasted with a single extended duration spike]

Once prolonged fasting is understood as an event, the question of repeatability becomes structurally important. Patterns are designed to recur. Events are not. A pattern is coherent because it can be repeated without losing its form. An event is coherent because it is bounded and discrete. These are not interchangeable properties. A prolonged fasting event may be metabolically intelligible as a single occurrence, but that does not make it structurally repeatable in the way a pattern is. Repeatability requires that the system be able to return to a sufficiently restored baseline before re-exposure. That assumption cannot be granted automatically after an event, because the event itself may have changed the baseline from which repetition would occur.

This is where recovery dependency becomes central. Events require recovery before reoccurrence because events are not simply episodes of exposure. They are periods of concentrated demand that reorganize internal conditions while they are underway. A prolonged fast is not metabolically neutral once it ends. Its effects do not disappear at the moment intake resumes, nor does the system automatically reset because the initiating condition has been removed. Input can return immediately. Restoration cannot. This distinction is not semantic. It reflects a basic asymmetry in biological systems: the factors that create strain can often be imposed faster than the factors required to fully resolve that strain can be re-established. An event ends when the condition ends. Recovery ends only when the system has re-stabilized. Those are not the same point in time.

That difference explains why prolonged fasting cannot be treated as though it were simply available for repetition whenever the initiating condition can be reproduced. The absence of intake can be repeated on command. The system state produced by the prior event cannot be assumed to have normalized on the same timeline. If a prolonged fasting event is repeated before recovery has sufficiently occurred, the second event does not land on an undisturbed system. It lands on a system still carrying unresolved constraints from the first. This is what transforms repetition from recurrence into compounding. The stress of the second event is not equivalent to the stress of the first because the baseline has changed. A system already narrowed by incomplete restoration has less flexibility, fewer reserves of tolerance, and a reduced capacity to distribute the burden of a new event cleanly.

This is why events do not reset automatically. Reset is often imagined as if biological systems simply return to default once the initiating challenge has stopped. But biological systems do not function like switches. They function through recovery processes that take time, depend on conditions, and may unfold unevenly across different domains. Energy handling may normalize on one timescale, endocrine signaling on another, substrate repletion on another, autonomic tone on another, and broader adaptive stability on another still. These layers do not necessarily re-synchronize immediately. A prolonged fasting event therefore cannot be evaluated only by whether the fasting condition has ended. It must also be evaluated by whether the system has meaningfully re-entered a state in which another major absence event would occur inside a restored rather than distorted architecture. Without that distinction, repetition begins to masquerade as resilience when it may simply be cumulative destabilization.

The problem becomes clearer when removal and recovery are viewed as structurally asymmetric processes. Input removal is immediate. Once intake stops, the initiating condition is fully imposed. There is no partial onset in the practical sense. The system is immediately placed into the absence condition and must begin regulating accordingly. Recovery does not operate with the same immediacy. Recovery is delayed, conditional, and distributed across time. It depends on what was depleted, what was suppressed, what was altered, and whether the environment after the event actually permits meaningful restoration. This creates an important imbalance: the system can be moved into stress faster than it can be returned from it. That imbalance is one of the defining features of event-based stress.

This asymmetry matters because it means that removal scales quickly while recovery does not scale proportionally. A longer event does not simply require a proportionally longer recovery in a neat one-to-one sense. The relationship is more complex because duration changes what must be recovered. A modest disturbance may require limited restoration. A more extended disturbance may involve multiple layers of adaptation, compensation, and downstream reorganization that do not reverse at the same rate they were induced. This is why event repetition becomes structurally unstable when it is interpreted through a pattern mindset. A pattern assumes recurrence with preserved recoverability. An event repeated under unresolved conditions violates that assumption. The system is asked to absorb repeated removal while recovery remains incomplete, and that mismatch progressively narrows the range within which stable regulation can still occur.

This narrowing is not always dramatic in presentation, which is one reason it is often overlooked. Systems can remain functional while carrying unresolved strain. Compensation can preserve apparent continuity even when internal margins are shrinking. But functional continuity is not the same as restored symmetry. A system can still operate while relying on increasingly constrained adaptation. The danger in treating prolonged fasting as a repeatable cycle is that the visible ability to continue can be mistaken for evidence of clean reset. It is possible for a system to remain active, responsive, and outwardly coherent while internally becoming less capable of tolerating further disturbance without distortion. This is one of the central structural errors corrected by the event-based model. Repetition alone does not prove repeatability. It may instead reveal a growing dependency on compensation.

Seen clearly, prolonged fasting belongs to a different temporal class than intermittent fasting. Intermittent fasting is organized around return. Prolonged fasting is organized around extension. Intermittent fasting derives its coherence from recurrence. Prolonged fasting derives its meaning from duration and the stress continuity that duration imposes. Intermittent fasting remains intelligible as a pattern because the organism repeatedly moves through bounded oscillations. Prolonged fasting becomes intelligible only when it is treated as a discrete event whose internal demands continue evolving while the event remains in progress and whose consequences do not automatically resolve once the initiating condition ends. These are not small variations inside the same framework. They are different structural categories.

That distinction matters because temporal architecture determines system burden more reliably than surface similarity. Both intermittent and prolonged fasting involve absence of intake, but the organism is not responding to a concept. It is responding to a timeline. Whether stress remains rhythmic or becomes continuous, whether restoration is built into the architecture or deferred until after the event, whether the system is oscillating or enduring—these are the variables that determine what kind of metabolic structure is actually being imposed. Once prolonged fasting is removed from the language of repeatable rhythm and placed into the framework of discrete event stress, its logic becomes much easier to understand. It is not a stronger pattern. It is a different system form, governed by different constraints, dependent on different recovery assumptions, and defined by a temporal asymmetry that cannot be ignored without distorting the phenomenon itself.

Magnitude

The most common structural error in thinking about prolonged fasting is not simply that it is treated casually, but that it is treated linearly. Linear thinking assumes that if a condition is tolerable at one duration, then a longer duration is merely more of the same condition in proportion to the time added. This assumption works reasonably well in systems where input and output remain approximately proportional across the relevant range. In a linear system, doubling an input roughly doubles its effect. Small changes remain small, larger changes remain predictably larger, and the system response is stable enough that extension can be understood as accumulation without transformation. That is not how biological stress behaves under sustained deprivation. Biological systems are not passive receivers of duration. They are adaptive, compensatory, threshold-sensitive systems whose internal state changes as exposure continues. This means that duration is not simply an additive variable. It is a transforming variable.

That distinction matters because the body does not encounter time as an empty container through which a fasting state is passively extended. It encounters time as a sequence of changing internal conditions. Each additional interval of absence is imposed on a system already altered by the intervals that came before it. As a result, the same extension in duration does not produce the same effect at every stage. A later increment of time is not metabolically equivalent to an earlier increment because it lands on a different system state. This is the defining feature of nonlinear stress scaling. The response curve bends. Magnitude increases disproportionately as duration extends because duration changes what the system must now manage. Stress is therefore not best measured by elapsed time alone, but by how elapsed time reshapes internal demands, reserve access, compensatory behavior, and the conditions under which regulation must continue.

A linear interpretation of fasting duration assumes that a system remains qualitatively similar while the clock advances. It assumes that the first phase and the later phase of a prolonged fast belong to the same stress class and differ only in degree. But once a system is understood dynamically, that assumption becomes difficult to sustain. Biological systems respond to prolonged absence through staged adaptation, reserve reallocation, signaling shifts, and progressive narrowing of available flexibility. These changes mean that later duration does not simply stack on top of earlier duration in a neat arithmetic way. Instead, the significance of time changes as the organism moves deeper into the event. Early exposure and later exposure are linked, but they are not interchangeable. The internal landscape evolves. The same fasting condition is still present at the surface, but the system operating beneath that condition is no longer the same system that began the event.

This is why biological systems under sustained stress are nonlinear by structure rather than by exception. They contain thresholds, redundancies, compensations, and reserve hierarchies that create curved rather than straight responses. A system may absorb an initial stress increment with relatively little visible disruption because multiple buffers remain available. But as those buffers are drawn upon, reweighted, or partially exhausted, later increments of the same stressor can begin producing disproportionately larger effects. This does not mean the system suddenly fails at a single dramatic moment. It means the conditions under which it is functioning become increasingly altered, and therefore increasingly sensitive to continued duration. What looks like a simple extension from the outside may represent a qualitatively different regulatory burden from the inside.

That burden becomes easier to understand once duration is linked not merely to exposure length, but to depletion depth. A prolonged fast does not simply last longer. It reaches deeper. Length describes how long a condition persists. Depth describes how far into the system that persistence reaches. This is a crucial distinction because biological reserves are not organized as a single uniform pool that can be accessed indefinitely without changing system behavior. They are layered. Some resources are available with relatively low structural disruption, while others become more relevant only as duration forces the organism to move beyond the more immediately accessible range. As a result, prolonged fasting is not only a matter of staying in a fasting condition for more time. It is a matter of progressing through increasingly deeper layers of adaptation and depletion.

Depth changes the meaning of the event because early-stage access and deeper-stage access are not metabolically equivalent. A system operating within the more superficial range of deprivation is often still managing absence with relatively broad flexibility. It can redistribute, compensate, and maintain coherence using reserves and regulatory adjustments that remain close to ordinary oscillatory function. But as duration extends, the system is required to solve the same problem with fewer low-cost options available. This is where depth becomes more important than duration alone. The body is no longer only continuing. It is continuing from further inside the stress event. That deeper position changes the conditions of regulation. The organism is not merely enduring more time. It is enduring more time from a more constrained state.

This layered view of depletion clarifies why later stages of prolonged fasting cannot be treated as straightforward extensions of earlier stages. What is being accessed, suppressed, conserved, or reweighted changes as the event deepens. Reserve use is not only quantitative but structural.

The system moves through different operating assumptions as duration continues, and those assumptions influence everything downstream: signaling, substrate allocation, tissue prioritization, autonomic behavior, and the eventual cost of restoration. A shallow depletion state and a deeper depletion state may both fall under the broad category of fasting, but they do not represent the same magnitude of system burden. The later state is not simply a more committed version of the earlier one. It is a different internal configuration, created by the cumulative consequences of time acting on a living system.

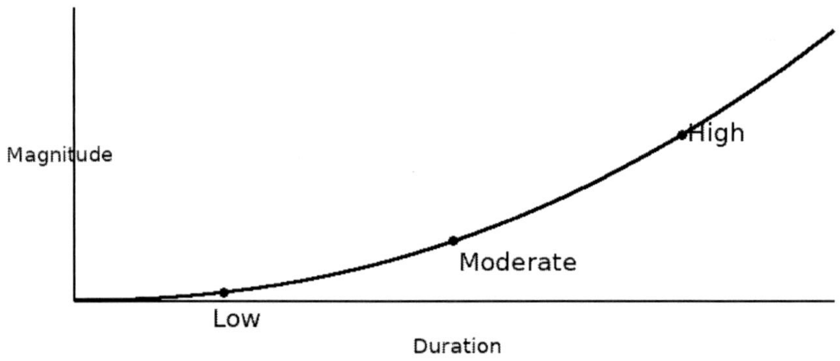

[Figure 3: Nonlinear magnitude curve—stress increasing slowly at first, then accelerating as duration extends]

Once depth is recognized, it becomes clear that magnitude cannot be inferred from duration alone in a simple one-to-one way. Magnitude is better understood as the total system burden created by how far the event has progressed into depletion, constraint, and compensatory adaptation. Two stress exposures of different durations are not separated only by the amount of time that has passed between them. They may differ in the type of internal work now required to maintain function. Early in a fasting event, the organism may still be operating inside a relatively broad adaptive field. Later, it may be functioning under narrower margins, more heavily dependent on compensation, with fewer inexpensive routes to preserve coherence. Magnitude therefore reflects not just how long the event has persisted, but how much system reorganization that persistence has already demanded.

This has direct implications for recovery. Recovery demand does not scale cleanly with duration because recovery is not tasked with reversing elapsed time. It is tasked with restoring what magnitude has altered. A prolonged fasting event of greater magnitude requires broader restoration because more domains have been recruited, reweighted, or constrained during the event. Recovery is not simply the mirror image of fasting. It is not the reverse playback of absence. It is a re-stabilization process that must account for how deeply the event has reorganized internal conditions. This is why the cost of a larger-magnitude event cannot be captured by asking only how long it lasted. What matters is how much depletion occurred, how many layers of regulation were involved, and how much adaptive bandwidth was consumed in the process.

Broader restoration is required because deeper events alter more than substrate availability. They affect the coordination between systems that must later re-synchronize. Energy balance may begin moving toward repletion quickly once input returns, but the organism is not restored merely because energy has re-entered the system. Recovery must also include restoration of signaling coherence, redistribution of substrate priorities, reduction of compensatory stress, and normalization of the broader regulatory environment that was progressively altered during the fast. The deeper the event, the more layers are implicated. This is why recovery demand scales with magnitude rather than simply duration. A longer event that reaches deeper layers of depletion may require disproportionately more re-stabilization than a shorter event, not because time itself is harder to reverse, but because the system state produced by that time is more altered.

This scaling problem becomes especially important when depletion and recovery are viewed together. Depletion can accelerate because the system is able to move rapidly into deeper stress territory once the easier layers of adaptation have been traversed. Recovery does not possess the same acceleration profile. Restoration is constrained by process, not just availability. Biological systems can often be pushed into strain faster than they can be brought back into coherence. This creates a structural mismatch. Magnitude can rise faster than recovery capacity can scale to meet it. That mismatch is not incidental. It is one of the defining reasons prolonged fasting should not be interpreted through simple duration arithmetic. A system can descend into deeper burden more quickly than it can climb back out of it, especially once multiple domains of regulation have been affected simultaneously.

This mismatch also explains why later-stage fasting stress is so frequently underestimated. When fasting is imagined as a smooth extension of an earlier condition, the mind tends to project early tolerability forward as though the same relationship will continue to hold. But nonlinear systems do not reward projection from shallow states into deeper ones. The apparent manageability of earlier exposure says relatively little about the burden of later exposure if the system state has materially changed in the meantime. The same outward condition can conceal a very different internal load once duration has moved the organism further into depletion and compensation. Stress may appear to be increasing slowly from the outside while, internally, the system is approaching a steeper segment of the curve where each additional interval imposes disproportionately more burden.

This is precisely where the concept of the oscillation envelope becomes necessary. If magnitude increases nonlinearly with duration, then duration is not only a question of how long the event persists but of how much pressure the event places on the boundaries within which stable oscillation can still occur. The oscillation envelope can be understood as the recoverable range inside which a system can absorb stress, adapt to it, and still return to coherent function without accumulating destabilizing distortion. Magnitude matters because it determines how forcefully the event pushes against that range. A low-magnitude disturbance may remain well within the envelope, leaving broad room for recovery and minimal pressure on system stability. A higher-magnitude event may drive the organism closer to the boundary where compensation intensifies, flexibility narrows, and the cost of maintaining coherence rises sharply.

This pressure is not simply a matter of whether the organism is "coping" in a visible sense. Systems can appear functional while operating near the edges of recoverability. The relevant issue is whether the internal strain imposed by the event remains containable inside the envelope without forcing the organism into increasingly unstable forms of compensation. As magnitude rises, the probability increases that the system will begin sacrificing flexibility in order to preserve continuity. This is one of the hidden costs of nonlinear scaling. A larger event does not only increase burden. It changes how burden must be managed. More of the system's adaptive bandwidth is consumed simply holding the structure together. This can preserve short-term continuity while silently reducing the room available for clean recovery.

The connection between magnitude and envelope pressure is therefore not abstract. It is the core of why prolonged fasting must be understood structurally rather than impressionistically. A longer duration is not simply "more fasting." It is a potentially steeper stress input acting on a system whose ability to remain inside a recoverable operating range depends on how much depth has already been traversed and how much compensatory strain has already accumulated. Magnitude tells us how much of the system's tolerance is being recruited. The oscillation envelope tells us whether that recruitment remains bounded or is beginning to threaten the conditions required for later restoration. Together, they reveal why prolonged fasting cannot be understood by treating time as a straight line.

Once duration is placed inside this nonlinear framework, the distinction between shallow and deep fasting exposure becomes much more than a matter of scale. It becomes a matter of system class. Shorter exposures that remain near the surface of the stress curve are not merely smaller versions of deeper events. They belong to different portions of the system response landscape. The body is operating under different assumptions, drawing on different layers of reserve, and generating different recovery demands depending on where duration has carried it. This is why duration extension cannot be interpreted as a simple additive move from one point to another. The curve bends because the system changes underneath it.

The magnitude scaling model makes that visible. Stress magnitude increases nonlinearly with duration because duration progressively alters the internal conditions under which further duration must be endured. Longer fasting does not merely mean more time without intake. It means greater depletion depth, broader system involvement, steeper recovery demand, and increased pressure against the limits of recoverable stability. Once that is understood, prolonged fasting can no longer be framed as a neutral extension of a familiar state. It must be understood as a nonlinear stress event whose burden grows by changing the system itself as it unfolds.

Thresholds

A nonlinear system is not fully understood until its transitions are understood. Magnitude scaling explains that stress does not increase proportionally with duration, but that alone does not explain why system behavior can seem stable for a period and then become qualitatively different without any obvious external change in condition. The missing concept is threshold. A threshold is not merely a point on a timeline or a convenient label applied after the fact. It is a boundary condition within a living system at which the governing behavior begins to change. Before a threshold is approached, a system may remain inside a range where adaptation is still managing the imposed stress with relative coherence. After that range is exceeded, the same stressor can begin producing different forms of strain, different compensatory demands, and different consequences for stability. The external condition may appear continuous, but the internal system is no longer operating under the same rules.

This is why prolonged fasting cannot be understood as a smooth, continuous gradient of increasing challenge. The absence of input may be continuous, but the organism's response to that absence is not infinitely smooth. Biological systems operate within tolerable ranges. They can absorb perturbation, redistribute load, and maintain coherence so long as the demands imposed remain inside those ranges. Once the underlying limits of a regulatory range begin to be exceeded, the nature of the system response changes. What had been adaptive becomes more constrained. What had been manageable through ordinary compensation becomes more dependent on increasingly narrow forms of regulation. Thresholds therefore separate not just degrees of stress, but different regulatory states. They mark transitions in how the organism is preserving continuity under the same surface condition.

A threshold is not arbitrary because it does not originate from a conceptual model imposed onto the body. It emerges from the body's own architecture. Every adaptive system contains operating ranges, reserve margins, and tolerance boundaries beyond which its previous mode of regulation becomes less reliable. These boundaries are not always fixed to a single visible marker, nor do they necessarily produce dramatic changes the moment they are crossed. But they are real in structural terms because systems cannot adapt indefinitely without changing the form of their adaptation. A threshold is the point at which accumulated stress, depletion, or compensatory demand begins to reorganize system behavior rather than merely intensify it. That is what makes threshold transitions essential to understanding prolonged fasting. The question is not simply how much stress is present, but whether the stress remains inside a regulatory range that still supports adaptive coherence.

This is also why thresholds are inseparable from state. A threshold is never crossed in abstraction. It is crossed by a system already in a particular condition, carrying a particular burden, with a particular set of reserves and compensatory options still available. The same duration of fasting does not represent the same threshold relationship in every context because duration interacts with the condition of the system that is experiencing it. Thresholds therefore should not be imagined as universal lines that all bodies approach identically. They are better understood as dynamic boundaries between different forms of system organization. On one side of the boundary, the organism is still operating inside a broadly adaptive state. Beyond it, regulation becomes more constrained, less flexible, and increasingly dependent on preserving function under narrowing margins.

That transition may be gradual or abrupt, but once it occurs, the system is no longer behaving as though nothing fundamental has changed.

To understand this clearly, it helps to distinguish between soft thresholds and hard thresholds. Not all system transitions occur with the same sharpness. A soft threshold is a boundary that is crossed gradually. The organism begins showing increasing strain, reduced flexibility, or growing dependence on compensation before any overt loss of coherence becomes obvious. The system is still functioning, but it is doing so under altered assumptions. Adaptation has not disappeared, yet it has become narrower, more conditional, and more costly. Soft thresholds are important because they account for the fact that destabilization often begins before it is recognized. A system does not need to collapse to have already changed categories. It may remain outwardly continuous while internally moving from stable adaptation into a more constrained and less resilient state.

This is often where prolonged fasting is most easily misread. If visible dysfunction is used as the only marker of meaningful change, then soft threshold crossings are likely to be missed entirely. But biological systems often reveal instability first through subtle narrowing rather than overt failure. A system can still preserve continuity while losing margin. It can still maintain output while becoming increasingly dependent on compensatory regulation. That is what defines a soft transition. The rules are changing before the appearance changes. In prolonged fasting, this matters because the organism may continue to appear metabolically coherent while already operating closer to the edge of its recoverable range. The absence of obvious disruption does not mean the same regulatory state is still intact.

A hard threshold is different in that the transition is sharper and the loss of stability more pronounced. This does not necessarily mean dramatic collapse in a visible sense, but it does mean that the system's prior mode of adaptation is no longer sufficient to preserve coherence without major distortion. A hard threshold marks a point at which the organism's tolerance for further perturbation has narrowed enough that continuation begins producing qualitatively different consequences. Compensation may become more extreme, flexibility may contract rapidly, and the system's ability to preserve internal organization may become more fragile. Hard thresholds are therefore not simply "worse" versions of soft thresholds. They represent a more decisive shift in system behavior, where the prior adaptive range has been meaningfully exceeded.

Most biological transitions begin as soft before they become hard. This is a crucial point because it means thresholds often unfold progressively. The system usually does not move from perfect adaptation to overt destabilization in a single instant. It moves through a transitional zone where strain accumulates, flexibility narrows, and regulatory behavior begins to change before instability becomes unmistakable. This progression matters because it reveals that threshold behavior is not only about endpoints. It is about trajectory. The body often enters destabilizing territory by degrees, with each increment of continued stress pushing it further from broad adaptability and deeper into constrained continuity. The fact that hard thresholds are often preceded by softer transitions does not make them less real. It makes them more biologically plausible.

This also explains why thresholds are frequently invisible from the outside. External continuity can conceal internal transition because the body is built to preserve outward function for as long as possible. Biological systems prioritize continuity. They compensate, redistribute, suppress, and narrow optionality in order to maintain core operation under strain. That means the visible surface of functioning often lags behind the underlying shift in system state. By the time overt instability is recognized, the threshold crossing may have occurred much earlier. This is one of the reasons prolonged fasting is so easily flattened into a simple question of duration or tolerance. If thresholds are assumed to exist only where obvious dysfunction begins, then the transitional zone where the system has already changed categories remains conceptually invisible.

Once thresholds are understood as dynamic boundaries rather than fixed external markers, variability becomes unavoidable. No threshold exists independently of the system that is approaching it. Baseline condition matters because the starting organization of the organism determines how much adaptive range is available before constraint begins to dominate. A system with broader reserve margins, more coherent oscillation, and greater recovery integrity will not encounter the same transition boundary in the same way as a system already carrying latent strain, narrowed flexibility, or incomplete restoration from prior stress. This does not mean that thresholds are arbitrary. It means they are conditional. They arise from the interaction between the stressor and the condition of the organism receiving it.

Recovery capacity is one of the most important variables in this regard because thresholds are not defined only by how much strain a system can tolerate in the moment, but by how much strain it can absorb without destabilizing its broader ability to restore afterward. A system with greater recovery capacity may remain adaptive deeper into a prolonged fasting event because its regulatory architecture can absorb and reverse perturbation more efficiently. A system with narrower recovery bandwidth may begin crossing into constrained or destabilizing states earlier, not necessarily because it is weaker in a simplistic sense, but because its available margin for maintaining coherent adaptation is smaller. Thresholds therefore depend not only on immediate tolerance, but on the broader architecture of recoverability.

Accumulated stress also shifts threshold location. A system carrying unresolved load from prior disturbances is not approaching a prolonged fasting event from neutral ground. It is already positioned closer to certain boundaries than a fully restored system would be. This is one of the most important reasons thresholds must be understood as dynamic rather than fixed. A threshold is not a static line waiting to be reached. It moves according to system condition. The organism's current state changes how much additional stress can be integrated before adaptation begins giving way to compensation and destabilization. The same external duration therefore does not produce the same internal consequence across individuals, or even within the same individual across different conditions. Time alone is not the threshold. Time interacting with state is.

This dynamic quality makes prolonged fasting especially difficult to interpret through universalized assumptions. If thresholds vary by baseline condition, recovery capacity, and accumulated stress, then the same fasting exposure can sit inside adaptive range for one system while pushing another into a transitional or unstable zone. That does not make the phenomenon subjective. It makes it biological. Living systems are stateful, not abstract. Their thresholds emerge from real differences in available margin, regulatory coherence, and prior load. To ignore that variability is to mistake surface sameness for structural equivalence. The absence of intake may be identical as an initiating condition, but the threshold relationship to that condition is not identical across all organisms.

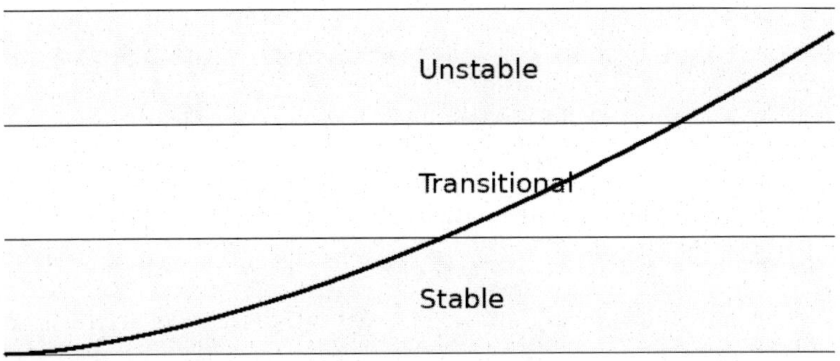

[Figure 4: Stable → Transitional → Unstable bands—layered zones showing progression from stability to instability as duration increases]

Once a threshold is crossed, the consequences are not limited to "more stress." Crossing a threshold changes the form of system behavior. One of the earliest consequences is loss of adaptive flexibility. Flexibility is the capacity of a system to redistribute load, preserve coherence, and still remain broadly recoverable under variation. When a threshold is exceeded, that flexibility narrows. The organism can still function, but it is doing so with fewer options. Regulation becomes more rigid, more constrained, and more dependent on preserving continuity through narrower pathways. This is a fundamental shift because a flexible system and a constrained system are not simply separated by degree. They operate under different rules of tolerance.

As flexibility narrows, compensation becomes more central. Compensation is not inherently pathological. It is one of the main ways biological systems preserve continuity under stress. But compensation becomes structurally important when it begins replacing broad adaptability rather than supporting it. Once a threshold is crossed, the system increasingly relies on compensation not as a temporary aid, but as the primary means of maintaining function. This creates a different kind of stability—one that is less robust, more conditional, and more vulnerable to additional perturbation. The organism may still appear stable, but the stability is now contingent on holding together under constrained conditions rather than operating inside a broad adaptive range.

This shift has direct implications for tolerance. Tolerance is not just the ability to continue. It is the width of the range within which continuation remains recoverable. Threshold crossing narrows that range. The system can often still absorb some further stress after a threshold has been exceeded, but it does so with a reduced margin for error and a greater likelihood that additional burden will produce disproportionate destabilization. This is why threshold transitions are so important in prolonged fasting. Once a threshold is crossed, continuation is no longer occurring under the same assumptions as before. The system is not simply enduring more of the same event. It is enduring the event from a different regulatory state, one in which further stress may carry qualitatively different consequences.

Destabilization, then, should not be understood as immediate collapse. Collapse is only one extreme form of destabilization, and often not the most relevant one for systems analysis. More often, destabilization is progressive. It appears as increasing rigidity, deeper reliance on compensation, reduced reversibility, and narrowing oscillatory range. The system continues, but it does so less freely and less cleanly. This is the more important meaning of threshold crossing in prolonged fasting. The event may remain outwardly continuous, but internally the organism has shifted from adaptation toward maintenance under constraint. That transition changes everything that follows. It changes what continued duration means, what recovery must later resolve, and how much further burden can be absorbed before the system moves into still more unstable territory.

The threshold transition model makes this visible. Prolonged fasting does not move along a single smooth gradient where each increment of duration simply adds more stress in a uniform way. It moves through ranges of system behavior separated by threshold boundaries. Within one range, adaptation may remain coherent. Beyond another, flexibility begins to narrow and compensation grows more central. Beyond another still, stability itself becomes increasingly fragile. These boundaries are not arbitrary, not fixed, and not always externally visible, but they are real because they reflect the changing relationship between imposed stress and the system's capacity to absorb it without reorganizing into a less stable form. Once this is understood, prolonged fasting can no longer be interpreted as a simple extension of absence. It must be understood as a threshold-governed stress event whose meaning changes when the system crosses from one regulatory state into another.

The Depletion Curve

A prolonged fast is often described as though it were a single metabolic state that simply persists for a longer period of time. That framing is structurally inaccurate because energy systems do not remain static under sustained absence of input. They transition. The relevant metabolic question is not merely whether the body is fasting, but what substrate configuration the body is operating under at a given stage of the event. Prolonged fasting is therefore better understood as a time-dependent progression across energy systems rather than a single unified condition. As duration extends, substrate reliance shifts, reserve access changes, and the organism moves through different regulatory states that are linked to one another but not metabolically identical. What appears externally as one uninterrupted event is internally a sequence of transitions.

That sequence matters because substrate use is not only about fuel availability. It is about how the organism is organizing itself around the changing constraints of continued absence. A shift in substrate use is not simply a swap from one fuel source to another. It reflects a broader reconfiguration of energy allocation, hormonal signaling, tissue prioritization, and metabolic assumptions. Each substrate phase therefore carries its own system logic. Glycogen dependence, increased fat mobilization, and later ketone reliance are not just biochemical waypoints along a smooth path. They represent distinct operating conditions inside the same event. The progression across them is one of the primary ways prolonged fasting becomes metabolically different from a short fasting interval. Time changes the substrate landscape, and the substrate landscape changes the system.

The first major substrate condition is glycogen dependence. Glycogen is the most immediately accessible stored carbohydrate reserve and serves as the organism's early buffer against the abrupt removal of external intake. It allows the system to preserve continuity when feeding stops by supplying glucose from internal stores rather than requiring an immediate and total reorganization of energy metabolism. In this sense, glycogen is not just stored fuel. It is a stabilizing bridge. It absorbs the initial shock of input removal and allows the organism to continue functioning inside a relatively familiar energetic architecture during the early phase of fasting. This makes it metabolically important not only because it can be used, but because its availability delays the need for deeper system transition.

That delay is structurally significant. Early-stage fasting remains glycogen-dependent not because the organism is incapable of using other substrates, but because glycogen provides the most immediate route for maintaining continuity while preserving as much short-term stability as possible. This means the initial phase of fasting is not yet defined by full adaptation to prolonged absence. It is defined by buffered transition. The system is still drawing from a relatively accessible reserve that permits continuity of glucose availability while the broader implications of continued absence are still unfolding. This is one reason the earliest phase of fasting is often misunderstood. It can feel metabolically smoother not because the event is simple, but because glycogen temporarily masks the deeper reorganization that prolonged absence will eventually require.

Glycogen depletion, then, is not merely a reduction in available energy. It is a transition trigger. As glycogen stores decline, the organism is pushed toward a different regulatory condition because the metabolic architecture supported by that reserve becomes progressively less sustainable. This is what makes glycogen depletion time-dependent in a meaningful systems sense. It is not just that a resource is being used up. It is that the depletion of that resource changes the conditions under which the rest of the event must proceed. The system is no longer operating with the same substrate assumptions it had at the outset. Once glycogen becomes less available, energy allocation, hormonal signaling, and substrate prioritization begin shifting under pressure. The organism is not simply continuing without food. It is continuing from a metabolically altered position.

This is the beginning of the depletion curve as a system phenomenon. The depletion curve does not describe only how much stored energy remains. It describes how the changing availability of specific substrates alters the organism's mode of regulation over time. Glycogen depletion is the first major bend in that curve because it initiates the move away from short-range continuity and into deeper reliance on internal reserve mobilization. It marks the transition from buffered absence toward more structurally consequential absence. This is why the early phase of prolonged fasting cannot be treated as a stable baseline for interpreting what follows. The system that begins the event with glycogen support is not the same system that continues after glycogen support has meaningfully declined. The event is already changing class from within.

As glycogen dependence diminishes, fat becomes the next major substrate. This shift is often described as though the body simply "switches to fat," but that phrasing is too clean to capture what is actually happening. Fat utilization is not just a replacement of one fuel source with another. It reflects a deeper movement into internal dependence. Once glycogen support becomes less available, the organism must increasingly rely on stored lipid reserves to maintain continuity across tissues and processes that still require ongoing energy provision. This changes the architecture of fasting in an important way. The event is no longer being buffered primarily by a relatively accessible short-range reserve. It is being carried more heavily by mobilization from deeper internal stores.

That shift matters because substrate allocation changes when fat becomes more central. Glycogen-supported fasting preserves a metabolic environment still relatively close to fed-state continuity, even though intake is absent. Increased fat utilization represents a more pronounced departure from that continuity. The system must now manage ongoing function through greater dependence on reserve release, transport, and tissue-specific allocation under sustained deprivation. This is not simply a more advanced form of the same state. It is a different energetic arrangement. The organism is no longer relying mainly on what was immediately available when feeding stopped. It is increasingly relying on what must be mobilized, distributed, and used from within.

This deeper internal dependence also changes how stress is experienced by the system. A condition supported by glycogen and a condition supported more heavily by fat are not metabolically equivalent, even if both are classified as fasting. The latter requires a broader commitment to internal resource use and a more sustained reorganization of substrate handling. That reorganization affects not just fuel access, but the overall logic of how the organism is preserving continuity under absence. The system becomes more dependent on what it can derive internally and less buffered by what was already prepared to carry it through the early phase. This is one reason prolonged fasting should never be understood as a static condition. The substrate base beneath it is moving.

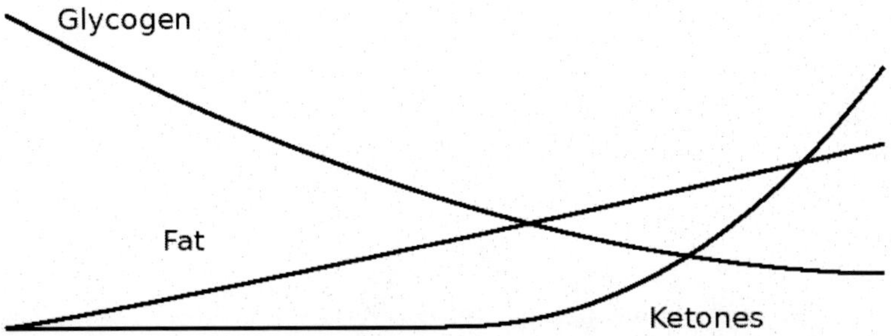

[Figure 5: Fuel transition curves—declining glycogen, rising fat utilization, delayed rise in ketone reliance over time]

As duration continues and depletion deepens further, ketones emerge as a later-stage adaptation. Their appearance is often interpreted in simplistic terms as proof of "being in a fasting state," but that collapses a dynamic process into a symbolic marker. Ketone reliance is more meaningfully understood as evidence that the organism has moved into a deeper stage of the event where the substrate configuration has changed again. Ketones are not the beginning of fasting. They are part of a later adaptive response to sustained depletion and ongoing absence. Their increasing relevance indicates that the system is no longer simply managing a short interruption in intake or operating mainly on the transitional support of glycogen decline and fat mobilization. It is functioning inside a more advanced configuration of substrate dependence.

This matters because ketone reliance represents more than fuel substitution. It reflects a different system arrangement. A body operating with increasing ketone contribution is functioning under metabolic assumptions that do not define the earlier phases of the event. It has moved further along the depletion curve and therefore further away from the conditions that existed at the outset. The substrate environment, the signaling environment, and the adaptive context are no longer the same. This is why later-stage fasting cannot be described accurately as just "more" of early fasting. The organism is not simply persisting longer in one condition. It is reorganizing into a different energetic state as duration continues to act on the system.

That later-stage configuration is important because it makes visible the deeper structural point of the chapter: substrate transitions define system states. Glycogen-supported fasting, increased fat dependence, and later ketone reliance are not merely different fuel options available inside one static state. They are distinct metabolic phases, each representing a different internal condition with different constraints and different forms of system burden. This is one of the most important reasons prolonged fasting must be interpreted as a time-dependent event. The same external behavior—continued absence of intake—contains internally shifting metabolic realities. Duration changes what the body is running on, and what the body is running on changes how the event must be regulated.

But even this progression can be misunderstood if it is imagined as smooth, automatic, or frictionless. Substrate transitions do not occur without cost. The body does not simply glide from one energy configuration into another with perfect efficiency and no intermediate instability. Every metabolic transition introduces a period in which one substrate system is becoming less available or less dominant before the next is fully established as an effective primary support. This creates lag. And lag is not a minor detail. It is one of the defining reasons prolonged fasting can carry system strain even in the absence of obvious failure. The transition itself can be metabolically expensive.

This is because depletion and adaptation are not synchronized processes. A reserve can begin declining before the organism has fully reorganized around what must replace it. Likewise, a substrate can become increasingly available before tissues and regulatory systems are fully operating in a way that uses it with maximal coherence. This gap between availability and effective utilization is where transition inefficiency lives. It is the metabolic interval in which the system is no longer being supported as fully by the earlier substrate condition, but has not yet fully stabilized inside the later one. The event is still moving forward, but the organism is carrying a temporary mismatch between what is being lost and what is not yet fully integrated.

That mismatch matters because it introduces temporary instability into the event itself. A prolonged fast is not only stressful because of depletion depth or threshold pressure. It is also stressful because the movement between substrate states requires reorganization, and reorganization is rarely free. The body must adjust allocation patterns, signaling assumptions, and tissue-level energy handling while the event continues. This means the system is being asked to adapt while still under ongoing absence. There is no pause in the event while the transition is completed. The transition occurs inside the stressor. That alone is enough to make the process metabolically more complex than a simple sequence of neatly ordered phases.

This helps explain why system behavior may fluctuate during substrate transitions even when the broad direction of adaptation is coherent. Fluctuation is not evidence that the system has no adaptive logic. It is evidence that transitions carry cost and that adaptation has a time structure of its own. A substrate becoming relevant in theory is not the same as that substrate being fully integrated into stable system use in practice. The body often requires a period of adjustment during which energetic continuity is preserved, but not preserved with the same efficiency that would exist once the new configuration is more fully established. This is what makes transition inefficiency such an important part of prolonged fasting physiology. It reveals that the movement between metabolic states is itself part of the burden of the event.

This burden also reinforces why each phase of the depletion curve must be understood as a distinct system state rather than a simple waypoint on a smooth line. Glycogen dependence, increased fat reliance, and later ketone integration are metabolically linked, but they are not interchangeable. Each phase carries a different relationship to reserve access, signaling environment, and adaptive burden. And each transition between phases introduces the possibility of friction, lag, and temporary instability. The body is not merely consuming stored fuel in a passive sequence. It is actively reorganizing its operating logic as duration extends. That reorganization is what makes prolonged fasting a dynamic physiological event rather than a static abstention condition.

Seen this way, the depletion curve is not just a fuel map. It is a systems map. It shows how duration changes the body by forcing transitions across substrate environments, each with its own constraints and each requiring its own form of adaptation. Glycogen depletion initiates the process by changing the meaning of continued absence. Increased fat utilization deepens internal dependence and alters substrate allocation. Ketone reliance marks a later-stage configuration that reflects a more advanced position inside the event. And none of these shifts occurs without transitional cost. The body moves through them not as a machine changing settings instantly, but as a living system reorganizing under pressure. That is what makes prolonged fasting metabolically dynamic: not simply that energy is absent, but that the conditions of energy use keep changing while the event continues.

Adaptive Shift

A prolonged fast does not begin in an adapted state. It begins as a disruption. The removal of input is not simply the subtraction of external energy; it is the removal of a central organizing force that structures metabolic timing, substrate distribution, signaling coherence, and recovery sequencing. Baseline physiology is built around recurring input, even when variability exists within that pattern. When input is withdrawn, the system does not seamlessly transition into a new stable mode. It is forced into immediate reorganization. This reorganization is not optional. It is required to preserve continuity of function under altered conditions. The organism must continue supplying tissues, maintaining regulatory coordination, and preserving core processes while the primary external driver of metabolic structure has been removed.

This initial phase is therefore reactive rather than optimized. The system is not yet expressing a refined adaptive state. It is responding to disruption using available mechanisms that can be mobilized quickly enough to prevent loss of continuity. Substrate shifts described in the depletion curve are part of this response, but they do not occur in isolation. They interact with changes in hormonal signaling, autonomic regulation, and allocation priorities. Early responses are shaped by immediacy. Glycogen is accessed not because it represents an ideal long-term strategy, but because it is accessible. Fat mobilization begins increasing not because the system has fully reorganized around lipid dependence, but because the decline of glycogen forces it to. These changes are driven by necessity rather than by equilibrium. The system is attempting to maintain function while the structure that previously supported that function is no longer present in its usual form.

This creates pressure across multiple domains simultaneously. The organism must begin resolving new constraints as they appear, often before earlier adjustments have fully stabilized. The removal of input therefore initiates a cascade of reorganization pressures that are not neatly sequenced. Substrate availability changes, signaling patterns shift, and energy distribution becomes more dependent on internal reserves, all while the system is still adapting to the loss of external supply. This is what defines the initial disruption phase. It is not a smooth entry into fasting. It is a forced departure from baseline organization, with the system attempting to preserve continuity through immediate, reactive adjustment. Stability is not yet present as a property. Continuity is being maintained through rapid reconfiguration under constraint.

From this disruption emerges the need for compensation. Compensation is the system's primary mechanism for preserving function when baseline conditions are no longer available. It is not a separate process layered onto metabolism. It is embedded within it. Compensation involves redistributing load, adjusting priorities, and modifying regulatory signals so that essential processes can continue even when the environment no longer supports them in the usual way. In prolonged fasting, compensation becomes central because the system must continue operating without the recurring reinforcement of input. Multiple systems adjust simultaneously to achieve this. Substrate mobilization increases, hormonal profiles shift, and tissue-level priorities are recalibrated. These adjustments are not independent. They are coordinated attempts to maintain continuity under conditions that would otherwise disrupt it.

Compensation does not remove stress. It reorganizes it. When one domain can no longer carry its usual share of system load, that load is transferred elsewhere. The organism is not eliminating the burden imposed by prolonged absence. It is redistributing that burden across systems capable of temporarily absorbing it. This redistribution allows the system to remain functional, but it introduces secondary strain. When fat becomes more central as a substrate, for example, this is not simply a clean substitution for glycogen. It requires changes in mobilization, transport, and utilization that carry their own regulatory demands. When hormonal signaling adjusts to support ongoing absence, those adjustments may preserve short-term coherence while altering other aspects of system behavior. Compensation is therefore necessary for continuity, but it is not neutral. It changes the internal balance of strain.

This is why compensation should be understood as a structural feature of adaptation rather than a sign of its completion. A system that is compensating is still resolving disruption. It has not returned to a baseline state, nor has it reached a stable endpoint. It has reorganized enough to continue functioning under altered conditions, but that functioning is supported by ongoing redistribution of burden. Compensation creates a form of provisional stability, but it does so by shifting constraints rather than eliminating them. The system is held together through coordinated adjustment, not through restoration of the original structure that preceded the disruption.

As compensation continues, the system may reach a phase of partial stabilization. This phase is often interpreted as evidence that adaptation has been completed, but that interpretation confuses appearance with structure. Partial stabilization occurs when compensatory mechanisms have aligned sufficiently to produce a more coherent and less visibly fluctuating system state. Substrate transitions may feel less abrupt, signaling patterns may appear more consistent, and the organism may seem to be operating with greater ease compared to the earlier phase of disruption. This does not mean the system has returned to baseline or resolved the underlying constraints. It means the system has reorganized into a configuration that can be maintained for a period under the current conditions.

This configuration is conditional. It depends on the continued effectiveness of compensatory adjustments and on the system remaining within the range where those adjustments can preserve coherence. The apparent stability of this phase is therefore not equivalent to restored equilibrium. It is a form of managed continuity. The organism is operating under altered assumptions about energy availability, signaling structure, and resource allocation. These assumptions are not neutral. They reflect the fact that the system is still in a state of sustained absence. Partial stabilization is what allows prolonged fasting to continue beyond the initial disruption phase without immediate collapse, but it does so by maintaining a reorganized state that differs from baseline in both structure and constraint.

This distinction becomes clearer when examining what has changed between baseline and this stabilized phase. The system is no longer oscillating around the same input-driven rhythm. It is not engaging the same balance of substrates under the same timing conditions. It is not operating with the same distribution of recovery and expenditure processes. Instead, it is functioning under a new configuration that has been assembled through compensation. This configuration may appear stable, but it is not free. It carries the accumulated effects of the adjustments required to reach it. The organism is maintaining coherence within a narrower and more conditional operating range.

That narrowing is what defines the fragility of adaptation. Adaptation under sustained deprivation is not an expansion of system capacity. It is a reorganization that allows continued function within constrained conditions. As duration extends, constraint pressure does not disappear. It increases. The system must continue managing absence while also sustaining the compensatory structure that has been built to support that absence. This dual burden reduces flexibility. A system that has adapted to prolonged fasting is often less able to absorb additional perturbation without destabilization because more of its adaptive bandwidth is already committed to maintaining its current state.

This fragility is not always immediately visible because the system can remain outwardly coherent while internally operating with reduced margin. The stability observed in the partially adapted phase is therefore conditional stability. It exists only within the bounds that the current configuration can support. If those bounds are approached or exceeded, the system may destabilize more quickly than it would have prior to the event, not because it has become weaker in an abstract sense, but because it has less available flexibility with which to respond. Adaptation has reallocated capacity toward preserving continuity under sustained absence, and that reallocation leaves less room for additional stress without further narrowing.

Continued duration intensifies this effect. As the event progresses, the system must maintain its reorganized state for longer periods, and the demands of doing so accumulate. Compensation does not resolve constraint; it manages it. Over time, the cost of maintaining compensation can increase, and the conditions under which the adapted state remains viable can become more restrictive. This is why adaptation cannot be equated with stability. A system can feel more stable after initial adjustment because the most abrupt transitions have been integrated, but that stability is not equivalent to a broad, resilient equilibrium. It is a narrower, more condition-dependent form of coherence.

This narrowing also changes how additional duration is experienced by the system. Once an adapted state has been established, further extension does not occur from the same starting point as the initial disruption phase. It occurs from within a configuration already shaped by compensation and constraint. This means that additional duration can exert disproportionate pressure relative to earlier stages because it is acting on a system with reduced flexibility. The apparent ease that sometimes accompanies partial stabilization can therefore be misleading if interpreted as evidence that the system is now operating under low strain. In structural terms, the system may be carrying a different kind of strain—one that is less visibly turbulent but more tightly bound to the limits of its current configuration.

Adaptation, then, is not a destination. It is an ongoing process of reconfiguration under sustained deprivation. It begins with disruption, proceeds through compensation, reaches a phase of partial stabilization, and remains inherently fragile as long as the conditions that required reorganization persist. Each stage builds on the previous one, not by eliminating its constraints, but by integrating them into a new system state. That state allows continuity, but it does so within narrower bounds and with greater dependence on the ongoing effectiveness of compensatory mechanisms. Prolonged fasting is therefore not a stable adaptive condition that the body enters and maintains unchanged. It is a dynamic reorganization that evolves over time, continuously shaped by the interaction between duration, constraint, and the system's capacity to preserve coherence under sustained absence.

Downregulation

Adaptation under sustained deprivation does not continue indefinitely as an open-ended expansion of system flexibility. At a certain point, the logic of the response changes. The system no longer attempts primarily to preserve continuity through broad adaptive reconfiguration alone. It begins to reduce itself. This is the essence of downregulation. A prolonged fast does not merely force the organism to reorganize around the absence of input. It also pushes the organism toward conservation. That shift matters because conservation is not the same thing as adaptation in its broader, more dynamic sense. Adaptation can still involve active redistribution, reorganization, and temporary compensation designed to maintain continuity. Downregulation reflects something narrower. It is a reduction response. The system begins decreasing output in order to reduce the burden of maintaining itself under conditions of ongoing resource limitation.

This reduction is not incidental. It is one of the central ways biological systems preserve survival when the energetic and signaling environment no longer supports the same level of activity. The body is not designed to maintain all forms of output equally under prolonged deprivation. It is designed to prioritize. When absence continues, the organism begins reallocating away from optional or expensive functions and toward preservation of core viability. This does not mean the system becomes more elegant, more optimized, or more metabolically refined in a celebratory sense. It means it becomes more constrained. The reduction of output is a protective contraction. It narrows what the system is willing to support in order to remain coherent for longer under conditions that no longer justify full expenditure.

The first and most fundamental expression of this contraction is metabolic rate reduction. Metabolic rate can be understood as the baseline level of system activity required to sustain ongoing physiological function. It reflects the organism's standing level of energetic throughput, not merely during visible movement or effort, but across the background operations that keep tissues, signaling, repair, and regulation active over time. Under sustained deprivation, this baseline becomes increasingly difficult to justify at its prior level because the assumptions that supported it—recurring input, replenishment, and broad recoverability—are no longer intact. The system therefore begins lowering the energetic cost of its own maintenance.

This reduction should not be misread as a sign that the organism has become more efficient in a positive or upgraded sense. Efficiency implies preserved function with less waste. Downregulation is different. It is not primarily about doing the same work better. It is about doing less work in order to preserve viability under constraint. The distinction matters because the surface appearance can be misleading. A lower-output system may appear quieter, leaner, or less demanding, but that does not mean it is performing the same functions with equal breadth at a lower cost. Often it is preserving only what is most essential while reducing support for what is less immediately necessary. That is not optimization. It is conservation under pressure.

This is the direct continuation of the adaptive reconfiguration described earlier. Adaptation allows the system to maintain continuity under disruption by reorganizing around the new conditions. But as deprivation continues, the cost of sustaining even that reorganized state begins to matter more. Compensation itself becomes expensive. The body cannot indefinitely maintain an active adaptive posture under continued absence without eventually reducing the burden of that posture. This is where adaptation gives way to downregulation. The system stops trying only to redistribute and begins to contract. Constraint, rather than flexibility, becomes the governing logic.

This contraction extends beyond substrate handling into signaling itself. Signaling systems are not separate from energy conservation. They are one of its main instruments. A prolonged fast changes not only what fuels are used, but how strongly, how frequently, and in what pattern the organism communicates internally about resource availability, priority, and action. Signaling intensity may therefore decrease under sustained deprivation because the system is no longer operating under conditions that support the same amplitude of regulatory activity. Hormonal and neural communication patterns that are appropriate in a more energetically permissive state may become progressively dampened as the organism shifts toward preservation rather than expansion.

This dampening should not be interpreted as neutral simplification. Signaling is not decorative. It is part of how the organism coordinates its own complexity. When signals become suppressed or altered, the system is not merely becoming calmer. It is becoming narrower in how it organizes itself. The intensity and richness of internal communication often reflect how many options a system is still supporting at once. Under prolonged deprivation, the organism begins to reduce those options. Some signals may become less prominent, some less variable, some more rigid, and some less responsive. The body is not only using less. It is saying less to itself, in a regulatory sense, because the conditions of broad responsiveness are no longer being maintained.

This matters because signaling suppression carries cost even when it serves a protective function. A reduced signal environment may help lower expenditure and preserve short-term continuity, but it can also diminish flexibility, narrow responsiveness, and reduce the system's ability to coordinate across multiple domains with the same richness it had under less constrained conditions. Suppression is therefore not simply a lower-noise version of the same state. It is a changed state. It reflects a shift from broad, dynamic regulation toward more conservative, more selective control. That may preserve survival, but it does so by limiting what the system is trying to sustain.

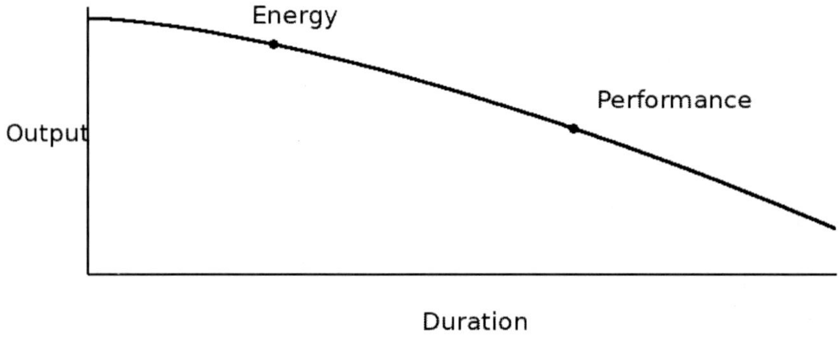

[Figure 6: Output decline curve—gradual reduction in system output as duration increases]

Once signaling begins narrowing, output reduction becomes more visible at the level of whole-system behavior. Output is not limited to overt movement or obvious effort. It includes physical activity, cognitive engagement, repair processes, regulatory responsiveness, and the general breadth of functions the organism is willing and able to sustain. Under prolonged deprivation, the system begins reducing these outputs because continued high activity across all domains would be incompatible with the logic of conservation. The body does not attempt to preserve every form of performance equally. It protects core continuity by selectively lowering what is less essential to immediate survival.

This selective reduction is often misunderstood because reduced output can appear orderly. A system that is doing less may seem more stable than a system in visible flux. But lower output is not evidence that the system has become more capable. It often means the system has become more selective. The organism is narrowing its operational field. It continues what must continue, but it does not maintain the same readiness to support optional, expansive, or high-cost forms of activity. This narrowing is what makes downregulation a structural response rather than a superficial one. It changes not only what the body is doing, but what the body is willing to keep doing under sustained absence.

At the physical level, this may manifest as a reduction in expendable activity, spontaneous output, or energetic willingness to sustain broad effort. At the cognitive level, it may appear as reduced range, reduced drive, or narrowed bandwidth for maintaining high-order engagement over time. At the regulatory level, it can mean slower responsiveness, lower adaptive amplitude, or a more restricted range of system movement. These are not separate phenomena. They are different expressions of the same conservation logic. The body is preserving core function by reducing the cost of everything that surrounds it. It is not shutting down indiscriminately. It is contracting strategically.

This strategic contraction is precisely what makes downregulation both protective and limiting. It preserves the most essential functions by reducing support for the less essential, but in doing so it also narrows system capability. Capability is not only about whether the body can continue existing. It is about how much complexity, responsiveness, and variation it can sustain while doing so. A downregulated system may remain coherent, but it is coherent within a smaller operational envelope. It is less metabolically permissive, less behaviorally expansive, and often less tolerant of additional demands. This is why downregulation should not be interpreted as a benign settling into a simpler state. It is a reduction in what the organism is prepared to support.

This reduction becomes even more important when considering what happens after the fasting event ends. Downregulation does not necessarily disappear the moment input returns. One of the defining features of prolonged conservation responses is that they can persist beyond the condition that initiated them. The body does not instantly restore prior output simply because external resources have reappeared. The system has already reorganized around scarcity. It has already adjusted expectations downward, narrowed operational range, and altered the relationship between expenditure and preservation. Those changes do not reverse automatically. Restoration must occur, and restoration is not instantaneous.

This persistence matters because it reveals that downregulation is not just an in-event phenomenon. It alters the recovery landscape. A system that has spent extended time reducing output to preserve viability may not immediately trust the return of input as a sufficient reason to restore all prior levels of expenditure, signaling intensity, and functional breadth. Recovery therefore involves more than replenishment. It requires the system to re-expand from a contracted state, and that re-expansion may lag behind the end of deprivation. The organism may continue behaving conservatively even after the external condition has changed because the internal logic of conservation has not yet been fully unwound.

This is one of the reasons prolonged fasting must be understood in terms of structural cost rather than only acute experience. A system that has downregulated has not simply paused until resources return. It has changed its operating assumptions. Those assumptions can continue shaping output, recovery dynamics, and tolerance for subsequent stress even after the fast itself has ended. This does not imply irreversibility as a universal rule. It means that conservation leaves a residue. The body carries forward the logic of what it had to become in order to endure the event, and that logic may continue influencing function until sufficient restoration and re-stabilization have occurred.

The deeper implication is that conservation introduces a form of inertia. Systems that contract under sustained deprivation do not always rebound symmetrically. Reduction can occur faster than re-expansion. The organism can move into a lower-output state relatively efficiently because survival depends on it. Returning to a broader, more permissive state may require more time, more conditions, and more internal confidence that the environment has actually changed in a stable way. This asymmetry reinforces the core point of the chapter: reduced output is not evidence of improved efficiency. It is evidence of a system preserving itself by becoming less expensive to run.

Seen clearly, downregulation is not a side effect of prolonged fasting. It is one of its central structural consequences. The system reduces output because continued deprivation changes what can be supported. Metabolic rate declines not as an optimization strategy, but as a conservation response. Signaling becomes more selective and dampened because broad regulation is costly. Physical, cognitive, and regulatory outputs narrow because preserving core function requires sacrificing breadth. And these changes do not necessarily disappear when the event ends, because the logic of conservation can persist beyond the condition that triggered it. Prolonged fasting therefore does not only reveal how the body adapts. It reveals what the body is willing to reduce in order to continue.

Cumulative Load

Prolonged fasting is often interpreted through whatever feature appears most visible at a given moment: lack of intake, substrate transition, subjective ease or strain, hormonal change, or reduced output. That way of reading the event fragments what is structurally continuous. A prolonged fast is not governed by isolated stressors appearing one at a time and then disappearing cleanly before the next emerges. It is governed by cumulative load. The system does not encounter energy limitation, altered signaling, reduced recovery latitude, and adaptive strain as separate episodes with independent meanings. It carries them together. This is why the concept of constraint must be defined clearly. A constraint is not merely a difficulty or a symptom of stress. It is a limit imposed on system function under sustained deprivation. It narrows what the organism can do, how flexibly it can regulate, and how broadly it can maintain coherence while the event continues.

Constraints are structural because they alter the operating conditions of the system rather than appearing as incidental side effects around an otherwise unchanged baseline. Energy limitation is a constraint because it reduces the availability of incoming substrate and forces the organism into deeper dependence on internal reserve allocation. Signaling disruption is a constraint because it changes the informational environment through which the body coordinates timing, priority, and response. Recovery limitation is a constraint because the system remains inside an ongoing event that delays or compresses the restoration processes needed to reverse accumulating burden. Adaptive strain is a constraint because the compensations that preserve continuity under deprivation are themselves costly, narrowing flexibility over time and making later regulation more dependent on maintaining altered conditions. These are not separate topics loosely associated with fasting. They are the structural terms under which prolonged fasting unfolds.

Once constraints are recognized in this way, the logic of the event changes. The question is no longer what single mechanism is dominant at a particular moment, but how multiple limits are shaping system behavior simultaneously and across time. Prolonged fasting is not difficult to understand because its individual components are obscure. It becomes difficult when those components are treated as though each one begins and ends in isolation. In reality, the organism does not stop being energy constrained when signaling becomes more altered. It does not stop carrying substrate transition costs when downregulation begins. It does not stop needing recovery merely because partial stabilization has emerged. Prior strain remains active as later strain develops. This is why stress compounds rather than resets. The system's burden is defined less by the existence of any one input than by the way unresolved constraints continue to shape the background into which further demands are introduced.

This cumulative logic becomes especially important in prolonged fasting because continued absence prevents the full clearing of earlier burdens while later burdens are still forming. Input withdrawal initiates the first layer of constraint by removing external supply and forcing reorganization around internal resources. That shift then changes signaling conditions, because the absence of intake is also the absence of recurring metabolic cues that ordinarily structure the system. As duration continues, recovery becomes less available as an active process because the event itself is still underway and repletion remains deferred. Adaptation begins to hold the structure together, but the cost of adaptation becomes a new source of strain rather than a replacement for the old one.

Downregulation reduces output in order to conserve resources, but this too narrows capability and therefore becomes another form of limitation. The system is not moving from one issue to the next. It is carrying forward each unresolved layer into a more densely constrained state.

This is the layering effect. Constraints do not occur independently, and they do not arrive as replacements for one another. They stack. Each new limitation is imposed onto a system already altered by the limitations that came before it. This means the organism never re-enters a clean baseline while the fast continues. The body at a later stage is not simply the same body with an additional challenge attached. It is a body whose operating state has already been redefined by prior constraint. Energy limitation may have altered substrate allocation. That altered substrate allocation may have changed signaling behavior. Those signaling changes may have affected recovery capacity and adaptive responsiveness. As each layer persists, it becomes part of the background condition for the next. The total burden of the event therefore cannot be measured by looking only at the newest stressor. It must be understood as the accumulated weight of all active limitations still being carried.

This stacking process is what makes prolonged fasting structurally different from a discrete and quickly resolved perturbation. In a short-lived event, the system may absorb a demand and return toward baseline before the next meaningful layer of strain develops. In prolonged fasting, continued deprivation prevents that return. Early burdens are not cleared; they are incorporated into the system's ongoing state. The organism continues forward not from restoration, but from unresolved adjustment. This is one of the central reasons that later-stage fasting cannot be interpreted by projecting forward from the earlier phase. The later phase includes the earlier phase within it, plus what has been added since. The system therefore becomes increasingly historical. What matters is not only what is happening now, but what the system is still carrying from before.

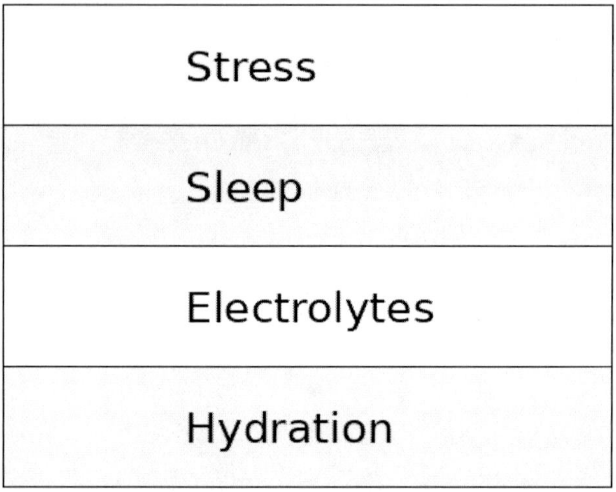

[Figure 7: Foundational Stability Stack—core layers that support system function and interact as a structured base.]

The importance of stacking becomes even clearer once it is understood that accumulated constraints do not simply sit beside one another as a passive collection. They interact. A constrained system is not a ledger of separate deficits. It is a network of overlapping pressures in which one limitation can alter the effect of another. Energy limitation, for example, does not only reduce available substrate. It can intensify signaling disruption by changing the conditions under which regulatory communication must occur. Signaling disruption does not only affect coordination in isolation; it can make recovery less efficient by weakening the coherence of the processes that would ordinarily support restoration. Reduced recovery capacity does not merely prolong burden; it can deepen adaptive strain by forcing compensation to remain active longer than would otherwise be necessary. Each constraint therefore has both its own weight and its effect on the others.

This interaction matters because it means cumulative load is not purely additive. If constraints simply stacked without influencing one another, the burden of prolonged fasting would still be substantial, but it would be easier to model as a sum of discrete parts. Biological systems are more complicated than that. Constraints modify the conditions under which other constraints are experienced. The presence of one limitation can increase the impact of another by reducing flexibility, narrowing response options, or weakening the system's ability to distribute load effectively. Combined burden therefore rises nonlinearly. The total weight of overlapping constraints is often greater than the simple total of each one measured independently because their interaction changes how the system behaves.

This is one of the reasons prolonged fasting becomes more complex as it continues. Complexity here does not mean richness or sophistication in a positive sense. It means that the system is increasingly governed by interacting limitations rather than by broad regulatory freedom. As constraints overlap, the organism's behavior can no longer be understood by isolating single variables and drawing direct conclusions from each. The same signaling shift means something different in a system with broad recovery margin than in a system already energy constrained and downregulated. The same reduction in output means something different in a system that still retains wide adaptive flexibility than in one already pressed by accumulated depletion and delayed restoration. Context becomes more decisive as constraints overlap because each new layer is filtered through the altered structure produced by the earlier layers.

This overlapping burden also changes the meaning of compensation. In an early phase of deprivation, compensation may preserve continuity with relatively broad effectiveness because the number of active constraints is still limited. As cumulative load grows, compensation itself becomes more difficult. The organism is not simply redistributing one burden at a time; it is attempting to coordinate around multiple interacting burdens whose combined effects are increasingly nonlinear. Compensation must now manage energy limitation while preserving signaling coherence, while containing adaptive strain, while operating inside reduced output, while awaiting recovery that cannot yet fully occur. The more these constraints overlap, the less freedom the system has to resolve any one of them cleanly. What once looked like flexible adaptation can harden into constrained maintenance.

This is the point at which cumulative load begins generating threshold pressure. Thresholds do not appear spontaneously. They emerge because accumulation changes the relationship between ongoing stress and the system's remaining capacity to absorb it. A threshold is crossed when the burden being carried exceeds the range within which adaptation can still preserve coherent function without reorganizing into a more unstable state. Cumulative load drives this process because it increases burden even in the absence of any dramatic new event. The system may appear to be continuing under the same general condition, but the actual internal cost of that continuation rises as unresolved constraints stack and interact. What changes is not only the amount of strain, but the amount of available margin left to contain it.

This is why threshold transitions are best understood as accumulation-driven rather than event-driven in any narrow sense. A system rarely becomes destabilized because of one isolated variable detached from its context. More often, destabilization occurs because cumulative burden has progressively narrowed the organism's room to regulate cleanly. A new layer of strain may appear to trigger the visible change, but the trigger only matters because the system was already carrying unresolved load that had reduced its adaptive range. In prolonged fasting, later-stage instability is therefore not best described as a sudden departure from an otherwise stable process. It is the outcome of accumulated pressure reaching a point at which the prior mode of adaptation is no longer sufficient.

This helps explain why system behavior can change qualitatively even when no single variable appears extreme on its own. A body can become more fragile not because any one component has reached an obvious maximum, but because multiple moderate constraints have accumulated and begun amplifying one another. Energy limitation may still be manageable in isolation. Altered signaling may still be manageable in isolation. Reduced recovery capacity may still be manageable in isolation. Lowered output may still be manageable in isolation. But isolation is the wrong frame. Once these pressures coexist, their combined effect can narrow tolerance more sharply than any one of them would suggest by itself. This is the practical meaning of accumulation threshold. The system is pushed toward instability not only by the size of individual loads, but by the density of unresolved load carried at once.

The cumulative model also clarifies why prolonged fasting cannot be understood by looking only at what is newly changing. Later stages of the event are shaped by inherited burden. The system carries forward substrate depletion, adaptive cost, downregulatory contraction, and deferred recovery into each successive phase. This continuity of burden is what gives prolonged fasting its compounding character. Stress is not applied to a fresh organism at each moment. It is applied to an organism increasingly shaped by its own prior adjustments. The body is therefore not only responding to current deprivation. It is responding to deprivation from within a structure already altered by the history of that deprivation.

Seen clearly, cumulative load is the integrating principle that links the earlier chapters into one continuous system argument. Substrate transition matters because it introduces energy constraint and changing fuel conditions. Adaptive reconfiguration matters because the system must reorganize under that stress. Downregulation matters because conservation becomes necessary as burden continues. But none of these mechanisms reaches its full meaning until they are understood as cumulative rather than isolated. Prolonged fasting is governed by the accumulation of active limitations that persist, overlap, and interact over time. Constraints shape behavior because they do not automatically resolve between intervals of strain. They are carried forward, incorporated into the system's current condition, and made more consequential by the addition of new burden on top of old.

The result is a metabolic event whose total cost cannot be inferred from any one mechanism alone. The burden of prolonged fasting is defined by accumulation. Each layer of unresolved constraint changes the conditions under which the next layer must be managed. Each interaction increases complexity and narrows flexibility. Each increment of carried strain pushes the organism closer to the point where adaptation becomes insufficient and threshold pressure becomes dominant. Stress, in this model, does not reset cleanly as the event continues. It compounds. And because it compounds, prolonged fasting must be understood not as a sequence of isolated challenges, but as a deepening architecture of cumulative load.

Hormonal Strain

Hormonal signaling is often treated as though it were a set of isolated chemical messages rising and falling independently in response to whatever condition happens to be present. That framing misses its actual structural role. Hormonal signaling is a coordination system. It does not merely report what is happening in the body; it helps organize what the body does next. Signals regulate timing, allocate resources, and synchronize behavior across tissues and systems that cannot coordinate effectively through substrate availability alone. This makes signaling central to any serious understanding of prolonged fasting. The organism is not simply burning stored fuel while hormones passively adjust in the background. It is attempting to maintain a coherent internal conversation under conditions of sustained deprivation. The quality of that conversation matters because biological regulation depends not only on having signals, but on having signals that remain interpretable, timed, and proportionate enough to support coordinated action.

Baseline hormonal signaling is dynamic, but it is not random. Coherent signaling does not mean static signaling. It means that fluctuations occur within structured ranges and maintain intelligible relationships to one another across time. Signals rise when certain conditions need to be emphasized, fall when those conditions recede, and interact in patterned ways that help the system distribute effort, conserve resources, and preserve stability. This dynamic coherence is one of the defining features of healthy regulation. A system remains adaptive not because every signal is quiet or symmetrical at every moment, but because signal behavior remains interpretable within the broader architecture of timing, feedback, and context. Hormonal activity is therefore best understood not as a collection of isolated outputs, but as a structured field of regulatory communication.

This baseline coherence becomes especially important in a state of prolonged deprivation because the body is then required to preserve internal coordination under increasingly constrained conditions. The same signal that is useful in one context can become misleading or destabilizing if its timing is distorted, its amplitude is narrowed or exaggerated, or its relationship to other signals becomes less reliable. Prolonged fasting therefore does not simply "change hormones" in a general sense. It changes the conditions under which signals are generated, interpreted, and integrated. The body is no longer coordinating under ordinary assumptions of recurring input, rhythmic repletion, and broad recovery latitude. It is coordinating under cumulative load. That matters because signaling quality depends on the stability of the environment through which signals are moving. When the environment becomes progressively constrained, signals do not merely become lower or higher. They become less clean.

This is where distortion begins. Distortion is not the same thing as reduction. A signal can be distorted without being absent, and a high signal can be just as distorted as a low one if its timing, amplitude, or context no longer align with the conditions it is supposed to help regulate. Under prolonged deprivation, hormonal signaling begins to shift because the system is no longer operating from a coordinated baseline. Energy limitation, altered substrate dependence, adaptive strain, and downregulation all change the internal conditions that shape signaling behavior. Signals are now being generated inside a system carrying cumulative burden. That burden changes what the organism is trying to prioritize and what it is still able to regulate cleanly. The result is not a neutral simplification of hormonal communication. It is progressive alteration.

Signal intensity is one of the first domains in which this alteration becomes visible. Some signals may become dampened as the organism shifts toward conservation and away from broad responsiveness. Others may become disproportionately emphasized because the system is attempting to preserve continuity through increasingly narrow priorities. But intensity alone is not the most important feature. A signal can be intense and still coherent if it occurs in the right context and within a reliable relational structure. What matters more is whether the signal remains accurate relative to the state it is meant to coordinate. Under prolonged fasting, that accuracy begins to erode because the system's internal conditions are no longer cleanly aligned. The organism is carrying unresolved load, and signals are now being produced from within that load rather than from within a broad, well-buffered regulatory range.

Timing is equally important. Hormonal signals do not only need to exist. They need to occur at the right moments relative to one another and relative to the organism's broader metabolic and regulatory state. Timing gives signals meaning. Under cumulative strain, however, timing can become less consistent. Signals that would ordinarily rise and fall in patterned relation to input, depletion, restoration, or oscillatory rhythm may begin to drift from those patterns. This does not require dramatic disorganization to matter. Even modest timing distortion can degrade coordination because the system depends on sequencing as much as intensity. A signal arriving at the wrong moment may not simply be less effective. It may create mismatch between what one part of the organism is being instructed to do and what another part is currently able to support.

Amplitude is another major dimension of distortion. Hormonal regulation depends not only on whether a signal is present, but on the range through which it can move. A signal whose amplitude has narrowed may still be active, but it has less room to modulate. A signal whose amplitude becomes exaggerated may preserve urgency at the cost of precision. In both cases, the problem is not only quantity. It is loss of proportion. Prolonged fasting changes the amplitude profile of signaling because cumulative deprivation alters both the baseline and the dynamic range within which signals operate. The organism may begin preserving a narrower band of hormonal movement, or it may show more volatile excursions in certain domains as compensatory demands intensify. Either way, the signal landscape becomes less proportionate to the system's broader coordination needs.

What matters most is that these distortions are not isolated defects within otherwise stable communication. They are changes in the quality of regulation itself. A distorted signal is not simply a weaker version of a baseline signal. It is a changed regulatory input moving through a changed system. That is why hormonal strain under prolonged fasting should not be understood as a simple reduction in endocrine activity. The body is not merely "slowing hormones down." It is producing and interpreting signals inside a context where the relationships that normally support coherence are under strain. The result is a signaling field that may still function, but does so with reduced clarity, altered proportion, and less reliable timing.

Once distortion appears in one domain, it does not remain there. Hormonal signaling is networked. Signals propagate because regulation in one part of the system depends on what is happening elsewhere. This is what makes hormonal strain so consequential. A distortion in one signaling stream can alter the conditions under which other signals are generated, received, or interpreted. The system does not contain these changes cleanly. It transmits them. A shift in one regulatory layer therefore becomes a change in the background conditions for multiple others. This is the propagation effect. Hormonal strain spreads not because signals are chaotic in a dramatic sense, but because coordinated systems are inherently interdependent.

That interdependence means that local distortion can create system-wide imprecision. A signal that has become less accurate in one domain may begin pulling resource allocation, timing behavior, or adaptive emphasis out of alignment in adjacent domains. The organism can still function under these conditions, but coordination becomes less exact. And when precision declines, the body increasingly relies on broader, more generalized forms of regulation rather than fine-tuned control. This shift is subtle but important. A system that is still regulating, but doing so with less precision, is not equivalent to a system that remains broadly coherent. It is functioning under reduced coordination fidelity. That reduction may preserve continuity, but it does so by tolerating more mismatch across the system.

This is one of the hidden costs of prolonged fasting as a cumulative event. The body does not merely lose resources over time. It loses signal clarity. And once clarity declines, the burden of regulation increases because the system must now preserve continuity with less precise internal communication. Signals still circulate, but their ability to coordinate timing, allocation, and adaptation across domains becomes progressively less reliable. This is not because the body has stopped regulating. It is because it is regulating under load. The signaling network is still active, but its outputs are now shaped by distortion and therefore carry distortion forward.

The problem deepens further when feedback is considered. Feedback loops are among the most important mechanisms by which biological systems maintain regulation. A feedback loop allows the organism to detect change, compare that change to its operating range, and adjust accordingly. This is how systems self-correct. They do not rely only on static settings or one-time commands. They remain stable by continuously updating themselves through reciprocal signaling. But this process depends on the signals within the loop remaining sufficiently accurate and interpretable. If signals are distorted, the feedback loop does not disappear, but its reliability degrades.

This degradation matters because self-correction is only as good as the information it receives and the precision with which it can respond. Under prolonged deprivation, feedback loops become less reliable not necessarily because the body stops attempting to regulate, but because the inputs driving correction are less clean. A system cannot correct precisely if the signals telling it what state it is in have become less accurate in timing, amplitude, or relational meaning. Correction mechanisms may still activate, but they do so from within a more ambiguous or distorted informational environment. This can lead to slower correction, partial correction, or compensatory responses that preserve continuity while leaving the underlying mismatch unresolved.

The result is a system that becomes less able to self-stabilize through its usual regulatory intelligence. It is still active, still signaling, still correcting where it can—but it is doing so with degraded fidelity. That degradation changes the quality of regulation in a profound way. A body with impaired feedback reliability may remain outwardly functional while internally becoming more dependent on compensation and less capable of fine-grained correction. This is how hormonal strain contributes to broader system instability. It does not need to produce dramatic endocrine collapse to matter. It only needs to reduce the precision with which the system can sense, interpret, and respond to its own changing state.

This is why hormonal signaling under prolonged fasting should never be imagined as a cleanly preserved control layer sitting above the metabolic event and managing it neutrally. It is inside the event. It is shaped by the same cumulative load that is affecting substrates, adaptation, downregulation, and recovery. The coordination system itself is under strain. Signals begin from a baseline of dynamic coherence, but as deprivation continues, that coherence becomes harder to maintain. Intensity shifts, timing drifts, amplitude changes, and the resulting distortion propagates across systems that depend on one another for coordinated regulation. Feedback loops remain active, but their reliability weakens as the signals moving through them become less precise.

Seen clearly, hormonal strain is not a side note to prolonged fasting. It is one of the central mechanisms by which cumulative load becomes system-wide. Signaling distortion changes not only what the body is doing, but how accurately the body can organize itself while doing it. And once the coordination layer begins to lose precision, the organism becomes less able to preserve coherent regulation under continued stress. That is what makes prolonged fasting more than a matter of fuel absence or energy conservation. It becomes, over time, a condition in which the body is increasingly required to govern itself through signals that no longer carry the same clarity they did at the outset.

Resource Conservation

A prolonged fast does not place the body into a neutral waiting state in which all functions are simply maintained at lower fuel availability until intake resumes. It places the system into a condition of prioritization. This is one of the most important structural shifts in prolonged deprivation and one of the easiest to miss when fasting is framed only in terms of fuel use or endurance. The body does not attempt to preserve all functions equally when resources are constrained. It reorganizes around hierarchy. Resource allocation becomes selective because the organism cannot sustain full breadth of output across all domains while external input remains absent and cumulative load continues to rise. The central logic of this phase is conservation, and conservation is not random reduction. It is ordered triage.

Resource allocation can be understood as prioritization under constraint. In a fully resourced or regularly replenished state, the system can support a wider range of functions simultaneously because it does not need to make hard decisions about what must be protected first and what can be reduced. Under prolonged deprivation, that flexibility narrows. The organism must begin treating its own functions as unequally urgent. This is not a sign of dysfunction in itself. It is the expected logic of a system attempting to preserve continuity when total available support no longer permits full-spectrum maintenance. The body does not become indifferent under constraint. It becomes more selective.

That selectivity reveals that physiological function is already hierarchical even when hierarchy is not obvious under ordinary conditions. Some processes are closer to immediate survival than others. Some outputs are expensive without being indispensable to short-term continuity. Some forms of system activity enhance range, resilience, and capacity, but can be reduced without immediately threatening core viability. Prolonged fasting forces this hierarchy into visibility because the system no longer has enough latitude to maintain every layer at equal intensity. Constraint exposes ranking. What had previously been sustained together must now be sorted.

This sorting is not primarily psychological or behavioral. It is structural. The organism begins organizing its internal economy around tiers of importance. Core survival processes are protected first because failure in those domains would compromise the continuity of the whole system. Processes that preserve baseline viability, internal regulation, and essential tissue function therefore remain comparatively defended. By contrast, outputs that are metabolically expensive, adaptively useful but nonessential in the short term, or not immediately necessary for continuity become more vulnerable to reduction. This is not because those functions cease to matter. It is because the system must now decide what it can afford to maintain fully while under sustained absence. Hierarchy is not an added feature of prolonged fasting. It becomes the governing logic of the event.

Once this hierarchy is in place, performance reduction becomes easier to interpret correctly. Reduced performance is often misunderstood because it is viewed through a framework that assumes the body should preserve ordinary output unless something has gone wrong. But in a conservation-dominant state, preserving ordinary output across all domains would itself be the wrong move. A system facing prolonged deprivation does not aim to demonstrate normal breadth. It aims to preserve continuity. That means output reduction is not evidence that the system has lost all logic. It is evidence that the system has changed its logic. It is no longer organized around maintaining wide capability. It is organized around protecting what matters most for ongoing viability.

This reduction is selective, not random. The body does not simply lower everything equally in an indiscriminate flattening of function. It scales back what is more expensive relative to its immediate necessity. This is why prolonged fasting should not be understood as a condition in which the organism becomes uniformly diminished across every domain at the same rate. It becomes more economical in a hierarchical sense. High-cost, nonessential, or expansion-oriented outputs are more likely to be reduced because they represent energy and regulatory expenditure that the system can no longer justify as easily under cumulative constraint. What is reduced is not arbitrary. It reflects the organism's internal accounting of what can be surrendered without collapsing the whole.

That internal accounting is often misread when performance is treated as though it were the body's default priority. Performance is not the body's default priority under deprivation. Survival is. This distinction matters because performance and survival are not synonymous. A system can remain alive, coordinated, and internally continuous while supporting a narrower range of output than it would under more permissive conditions. The body is not required to preserve its full functional breadth in order to preserve itself. In fact, under prolonged fasting, preserving that breadth would often conflict with the deeper logic of conservation. Reduced performance is therefore not necessarily a sign that the system has failed. It may be a sign that the system is making coherent decisions under pressure.

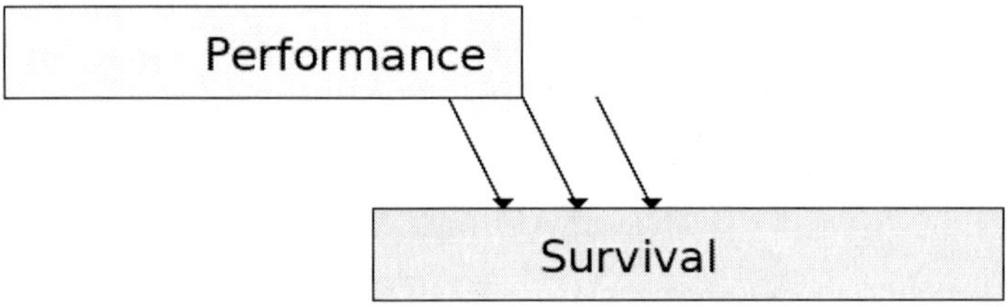

[Figure 8: Performance → survival shift—decline in performance output as survival prioritization increases under constraint]

This becomes clearer when survival preservation is examined directly. Survival, in systems terms, does not refer only to avoiding immediate collapse. It refers to maintaining the continuity of core viability across time. That continuity depends on preserving the processes and structures that allow the organism to remain metabolically organized, internally coordinated, and capable of eventual restoration once conditions improve. Under prolonged fasting, the body reallocates toward these functions because they form the non-negotiable center of the hierarchy. This does not mean they are untouched by deprivation. It means they are defended more strongly than functions the system judges to be more deferrable.

That defense requires reallocation. The body cannot preserve core viability under prolonged deprivation simply by "trying harder" to do everything at once. It must move resources toward continuity. This movement is what gives conservation its shape. Substrates, signaling, regulatory attention, and adaptive effort are increasingly directed toward maintaining what the organism deems indispensable. Other domains may still receive support, but not at the same priority level. The system is not operating from abundance and then trimming around the edges. It is operating from scarcity and building outward only as far as conditions allow. Survival preservation therefore drives not just a subset of responses, but the logic behind all major allocation decisions during prolonged fasting.

This logic also clarifies why the body may appear more stable in some ways even while becoming narrower overall. A system that has successfully shifted into survival-preserving prioritization can maintain coherence precisely because it has reduced what it is trying to sustain. It is not preserving stability by holding everything together equally. It is preserving stability by deciding what must be held together first. That can create the illusion of control or efficiency if the reduction in breadth is not recognized. But what looks like preserved order is often the result of a narrowed operating field. The organism has not transcended constraint. It has reorganized under it.

This is where tradeoffs become unavoidable. Conservation is not simply the art of doing less. It is the process of preserving one domain by withdrawing support from another. A tradeoff exists whenever the system cannot maintain full function across all domains simultaneously and must therefore choose which forms of support to defend. Under prolonged fasting, these tradeoffs are not optional and they are not merely side effects. They are the mechanism through which conservation operates. The body does not preserve survival first and then see what remains available for everything else by chance. It preserves survival by reducing what is less immediately defensible. That is the tradeoff.

This means that maintaining one layer of function often requires narrowing another. Supporting core continuity may require reducing broader responsiveness. Preserving essential regulation may require lowering the range of performance the system is willing to sustain. Maintaining internal stability may require tolerating reduced flexibility, reduced output, or delayed restoration in less urgent domains. These are not independent events. They are linked through the allocation hierarchy itself. Every act of preservation under deprivation is also an act of selective reduction somewhere else in the system. That is not a flaw in the model. It is the model.

The consequences of these tradeoffs become more significant over time because prioritization intensifies as cumulative load deepens. Early in deprivation, the organism may still retain enough flexibility to preserve a relatively broad field of function while beginning to shift toward conservation. As duration continues and constraints accumulate, however, the room for broad maintenance shrinks. The system becomes more aggressive in its hierarchy because it has fewer resources and less adaptive latitude with which to support nonessential range. This is why prolonged fasting does not merely reduce performance once and then settle into a neutral lower-output state. It can progressively narrow capability as the event continues. Tradeoffs accumulate because the cost of preserving core continuity increases the longer the event persists.

This progressive narrowing changes system behavior in a fundamental way. The organism is no longer attempting to preserve the widest possible range of function compatible with health under ordinary conditions. It is attempting to preserve the narrowest range of function necessary to remain coherent under prolonged absence. That is a different operating philosophy. And because it is different, expectations built around normal breadth, normal responsiveness, or normal flexibility become structurally misaligned with the actual logic of the state. The body is not behaving abnormally in the sense of losing all order. It is behaving according to a more restrictive order.

Seen clearly, this is what resource conservation really means. It does not mean that the body simply "slows down" in a vague or uniform way. It means the system begins making hierarchical decisions under constraint, protecting what is most necessary and reducing what is more expensive, more optional, or less immediately tied to continuity. Performance declines because performance is no longer the priority. Survival preservation drives allocation because survival, not expansion, is the central objective under prolonged deprivation. And because the system cannot fully preserve all domains at once, tradeoffs become the unavoidable currency through which that preservation is achieved.

This conservation logic changes how prolonged fasting must be understood. It is not a condition in which the body is trying to preserve normal function with fewer calories. It is a condition in which the body is actively reshaping itself around a narrower definition of what must be maintained. That reshaping is not random, and it is not merely the passive result of depletion. It is an ordered prioritization process driven by the need to preserve continuity under constraint. Once that is recognized, reduced capability, altered output, and narrowing function stop looking like disconnected costs and begin to appear as the coherent consequences of a system that has reorganized around survival first.

Refeeding Sensitivity

The end of a prolonged fast is often treated as a return point, as though the removal of deprivation restores the system to its prior condition. That interpretation assumes symmetry between entry and exit, as though the body can move into and out of absence with equal ease. The structure of prolonged fasting does not support that assumption. The organism does not exit the event from baseline. It exits from a constrained, adapted state shaped by cumulative load, substrate transition, downregulation, and altered signaling. This matters because the conditions under which input is reintroduced are not neutral. They are defined by the state of the system at the moment of re-entry, and that state is not one of broad readiness. It is one of narrowed capacity, altered allocation, and reduced operational range.

The low-input state created by prolonged fasting is not simply characterized by the absence of intake. It is characterized by the reorganization that absence has produced. Substrate handling has shifted across the depletion curve, with glycogen reduced, fat mobilization elevated, and ketone reliance reflecting a deeper stage of the event. Signaling has adapted to sustained deprivation, becoming narrower in amplitude, altered in timing, and shaped by cumulative strain rather than by rhythmic input. Output has been reduced through downregulation, narrowing the system's active range in order to conserve resources. These changes are not temporary surface adjustments that disappear when intake resumes. They define the structure from which re-entry begins. The organism is therefore not encountering food from a neutral metabolic position. It is encountering it from within a system that has reorganized around absence.

This reorganization creates a specific kind of mismatch when input returns. Input can be reintroduced immediately, but system recalibration cannot occur at the same rate. The processes that governed prolonged fasting—depletion, compensation, downregulation, and conservation—have shaped the organism into a state that is not immediately aligned with the conditions of renewed availability. When input reappears, it does not arrive into a system prepared to handle it with baseline coordination. It arrives into a system that has been operating under constrained assumptions about energy flow, signaling amplitude, and resource allocation. The result is not a smooth reversal of the fasting state. It is a condition of re-entry instability.

This instability begins with the speed of change. During prolonged fasting, the system moves into deprivation through a sequence of transitions that unfold over time: glycogen depletion, increasing fat reliance, signaling alteration, adaptive reconfiguration, and downregulation. These processes shape the system gradually, even when their effects accumulate. Reintroduction of input, by contrast, can occur abruptly. The organism can move from sustained absence to sudden availability in a way that compresses the transition back into a much shorter interval. This asymmetry creates a structural imbalance. The system is required to respond to a rapid increase in input with regulatory mechanisms that have been conditioned by a slower, cumulative process of deprivation.

That imbalance produces reintroduction shock. Shock, in this context, does not imply collapse or dramatic failure. It refers to the mismatch between what the system is receiving and what it is currently organized to handle. The organism has been operating under conditions where resource flow is limited, signaling is constrained, and output has been reduced to preserve continuity. When input returns, it introduces a level of substrate and signaling potential that exceeds what the system has recently been managing. The body is therefore presented with more than it is currently structured to integrate cleanly. This is not because the input itself is excessive in an abstract sense, but because the system receiving it has not yet recalibrated to a state that can distribute, interpret, and respond to it with baseline precision.

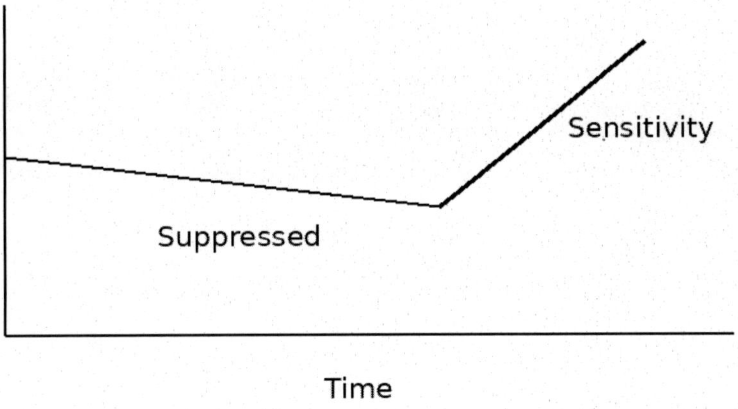

[Figure 9: Sensitivity spike—sharp increase in system response immediately following reintroduction of input]

From this mismatch emerges the sensitivity spike. Sensitivity, in this context, refers to the system's heightened responsiveness to input relative to its baseline state. This increased responsiveness is not a sign of improved efficiency or enhanced metabolic function. It is a reflection of instability. The system, having adapted to low-input conditions, is now reacting to reintroduced input with amplified responses because its regulatory architecture has not yet re-expanded to match the new environment. Signals that would normally operate within a balanced range may now produce exaggerated effects because the system's thresholds, amplitudes, and timing relationships have been altered during the fast.

This over-response can be understood as the system attempting to process input through a constrained framework. Signaling pathways that were narrowed or dampened during prolonged deprivation may now react more sharply when stimulated, not because they are functioning more effectively, but because they are operating outside their usual calibrated range. Substrate allocation mechanisms that were organized around internal reserve use may now encounter external availability in a way that disrupts their recent operating logic. The organism is not smoothly transitioning from one state to another. It is experiencing a temporary phase in which its responses are disproportionate to the input relative to baseline conditions.

This disproportionate responsiveness extends across multiple domains. Substrate handling may shift rapidly in response to renewed intake, but not necessarily with coordinated distribution across tissues. Hormonal signaling may become more reactive, with changes in intensity and timing that reflect the system's attempt to re-establish coordination under new conditions. Output may begin to increase, but without the full restoration of the regulatory environment that would normally support stable expansion. The system is therefore not simply "waking up" from a low-output state. It is navigating a period in which increased input meets an organism that is still structured for conservation.

This is why the sensitivity spike should not be interpreted as a beneficial surge in responsiveness. It does not represent a system that has become more capable. It represents a system that is temporarily less stable in how it processes input. The organism is attempting to integrate renewed availability while still carrying the constraints of the fasting state. This creates a window in which responses are more reactive, less proportionate, and less predictably aligned with baseline regulation. The system has not yet regained the buffering capacity, signaling coherence, and adaptive flexibility required to handle input with the same precision it had prior to the fast.

The duration and intensity of this sensitivity phase are shaped by the depth of the preceding event and the degree of constraint that has accumulated. A system that has undergone more extensive depletion, more pronounced downregulation, and greater signaling distortion will enter refeeding from a more constrained position. This does not change the underlying mechanism. It reinforces it. The greater the difference between the low-input state and the reintroduced input, the more pronounced the mismatch is likely to be. The system must then navigate a larger gap between its current configuration and the conditions it is being asked to process.

Stabilization, therefore, is not the automatic result of input returning. It is a process. The system must re-expand from its constrained state, and that re-expansion requires time and alignment across multiple domains. Substrate handling must shift from a depletion-driven configuration back toward a balanced distribution that integrates both internal and external sources. Signaling must regain coherence, with timing, amplitude, and relational patterns returning toward a state that supports coordinated regulation rather than reactive fluctuation. Output must increase, but in a way that is supported by restored capacity rather than driven by transient responsiveness. These changes cannot occur instantaneously because they depend on the system re-establishing conditions that were altered over the course of the fast.

This re-expansion is not simply the reversal of downregulation. It is the rebuilding of regulatory breadth. A system that has narrowed its operational range under prolonged deprivation cannot immediately widen that range without re-establishing the conditions that support it. Recovery processes that were deferred or compressed during the fast must resume. Signaling networks that were operating under constraint must regain their dynamic range. Substrate systems must move from a survival-oriented configuration to one that supports broader function. Each of these shifts requires coordination, and coordination requires time. The organism is not switching states. It is transitioning between them.

The instability of re-entry is therefore not an anomaly. It is a direct consequence of the asymmetry between how the system adapts to absence and how it responds to renewed presence. Prolonged fasting shapes the organism into a low-input, constrained configuration. Reintroduction of input challenges that configuration before it has had time to re-expand. The resulting sensitivity reflects the gap between those two states. The system is not immediately capable of processing input with baseline precision because it is not yet structurally aligned with the conditions of renewed availability.

Seen in this way, the end of a prolonged fast is not a simple endpoint. It is a transition phase defined by heightened responsiveness, reduced stability, and the need for re-synchronization across metabolic and regulatory domains. The organism must move from a condition of conservation and constraint back toward one of coordinated function under input. That movement is governed not by the presence of food alone, but by the system's ability to re-establish alignment between substrate handling, signaling, and output. Until that alignment is restored, the body remains in a state of re-entry instability, responding to input through a framework that was built for absence rather than for abundance.

Recovery Debt

A prolonged fast does not end when intake resumes. It ends when the system has been restored to a condition in which it can operate with broad coordination, stable signaling, and sufficient flexibility to absorb variation without distortion. That distinction matters because it separates the visible endpoint of deprivation from the actual completion of the event. The removal of absence is not the same as the restoration of capacity. Recovery is the process through which the system reconstitutes what was altered, constrained, or deferred during the fast. Without that process, the organism remains shaped by the event even after the initiating condition has been removed.

Recovery must therefore be defined precisely. It is not the reintroduction of input alone. It is the restoration of system capacity across multiple domains that were affected during prolonged deprivation. Substrate repletion is one component, but it does not fully define recovery. The organism must also re-establish signaling coherence, restore amplitude and timing relationships that were altered under constraint, and rebuild the broader coordination required for stable regulation. Output must expand from a narrowed, downregulated state, and flexibility must return so that the system can once again operate across a wider range without relying on compensatory narrowing. These changes are not instantaneous because they depend on processes that unfold over time. Recovery is therefore multi-domain and time-dependent. It cannot be reduced to the presence of input alone because input does not immediately restore the conditions that input normally supports.

This distinction reveals why recovery can be incomplete even when intake has resumed. The system may receive sufficient substrate to begin repletion, but signaling may remain distorted, output may remain reduced, and coordination may still reflect the constrained logic of the fasting state. Restoration requires alignment across these domains, not just replenishment in one. If alignment does not occur fully, the organism does not return to baseline. It transitions into a state that carries forward some portion of the prior constraint. This is the foundation of recovery debt.

Recovery debt is the gap between what depletion requires and what recovery delivers. It is not a metaphorical concept. It is the persistence of unresolved imbalance within the system. When a prolonged fast produces depletion across substrates, signaling, output, and adaptive capacity, it creates a set of restoration requirements. If those requirements are not fully met, the system does not return to its prior condition. It remains partially constrained. That constraint is not always visible as overt dysfunction. It can exist as reduced flexibility, altered signaling relationships, or narrowed adaptive bandwidth that persists beyond the end of the fast. The organism carries this residual burden forward because it has not been fully resolved.

This gap is important because it changes how subsequent stress is experienced. A system that has not fully recovered is not operating from the same baseline as one that has. It may appear functional, but its margin for absorbing additional load has been reduced. The organism is therefore not simply continuing from a neutral state. It is continuing from a state that has already been shaped by prior depletion and incomplete restoration. This is what makes recovery debt structural rather than incidental. It is not an abstract accounting error. It is an active condition of the system that influences how it will respond to future demands.

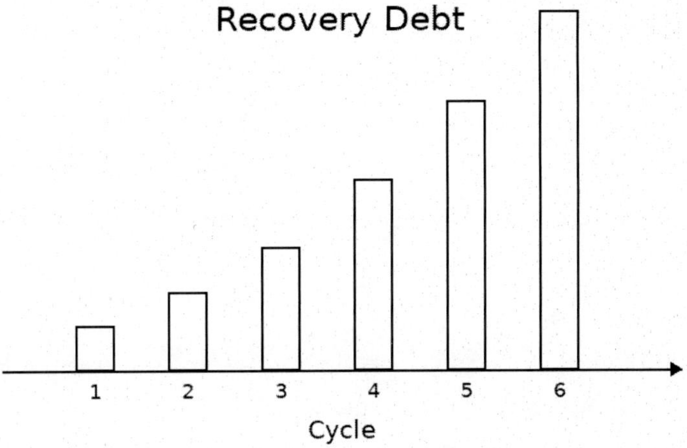

[Figure 10: Increasing recovery debt bars—stacked bars rising across repeated cycles where recovery is incomplete]

Once recovery debt is present, it does not remain static if the system is exposed to repeated cycles of depletion. It accumulates. Each new period of prolonged fasting imposes additional demands for restoration. If recovery remains incomplete, the new demands are layered onto a system that is already carrying unresolved strain. This creates a compounding effect. The organism does not begin each cycle from a fully restored baseline. It begins from a reduced one. That reduction may be subtle at first, but it alters the trajectory of subsequent exposure because the system has less flexibility, less reserve, and less capacity to absorb additional constraint without destabilization.

This accumulation changes the meaning of repetition. Repeated fasting events are often interpreted as discrete exposures, each of which can be evaluated independently. But if recovery debt is carried forward, the exposures are not independent. They are linked through the system's history. Each event begins from the state produced by the previous one, not from an abstract baseline. This means that the burden of a given event is not determined solely by its own duration or magnitude. It is determined by how much unresolved strain the system is already carrying when the event begins. The same external condition can therefore produce different internal consequences depending on the level of accumulated debt.

As this process continues, the baseline itself shifts. The organism becomes increasingly constrained, not because any single event has permanently altered it in isolation, but because the accumulation of unresolved strain has narrowed its operating range over time. Flexibility decreases, signaling coherence becomes more difficult to maintain, and recovery processes themselves may become less efficient because they are being initiated from a compromised state. The system is not simply oscillating between depletion and restoration. It is drifting toward a condition in which restoration is incomplete and depletion is layered onto that incompleteness.

This drift has direct implications for system stability. Stability depends on the ability to absorb perturbation and return to a coherent operating state without accumulating distortion. Recovery debt interferes with that process because it reduces the system's ability to fully resolve the effects of perturbation. When debt accumulates, thresholds are reached sooner. The system does not need to be pushed as far to enter a destabilizing state because it is already closer to its limits. Adaptation becomes less effective because it must operate within a narrower range and with less available capacity. Compensation may increase, but compensation itself is constrained by the same limitations that have been carried forward.

This creates a pattern in which instability becomes more likely even under conditions that might previously have been manageable. The organism is not becoming fragile in a superficial sense. It is becoming constrained in a structural sense. The range within which it can maintain coherent regulation has been reduced by the accumulation of unresolved strain. Each new demand therefore consumes a larger proportion of the remaining capacity, and the margin for maintaining stability without distortion continues to shrink.

This is why long-term system behavior cannot be understood by looking at individual fasting events in isolation. The outcome of a given event depends not only on its own characteristics, but on the accumulated recovery debt that precedes it. A system with minimal debt may absorb a period of prolonged fasting and restore itself with relatively little residual strain. A system carrying significant debt may experience the same event as a much greater burden because it begins from a reduced baseline and has less capacity to complete the recovery that the event requires. The difference is not in the event itself, but in the system that receives it.

Recovery debt therefore functions as a form of stored imbalance. It represents the portion of prior depletion that has not been resolved and continues to shape the system's current condition. This stored imbalance is not inert. It influences signaling, allocation, adaptability, and threshold behavior. It determines how quickly the system moves toward instability under new stress and how effectively it can return to coherence after that stress is removed. In this way, recovery debt becomes one of the central determinants of long-horizon system behavior.

Seen clearly, the end of a fast is not the end of its effects. The system carries forward whatever portion of the event has not been resolved through recovery. If recovery is complete, the organism returns to a baseline capable of supporting broad function and stable regulation. If recovery is incomplete, the organism remains partially constrained, and that constraint becomes part of the starting condition for whatever follows. Over time, repeated mismatches between depletion and recovery transform isolated events into a cumulative burden that reshapes system capacity itself.

This is the core of the recovery debt mechanism. Prolonged fasting creates depletion that must be matched by restoration. When restoration does not fully meet that requirement, the gap becomes debt. That debt accumulates when new cycles occur before prior ones have been fully resolved. And as debt accumulates, the system's ability to maintain stability, adapt effectively, and recover completely becomes progressively reduced. The organism is no longer responding to each event from a neutral baseline. It is responding from within a history of unresolved imbalance that continues to influence its behavior long after the visible absence has ended.

Rebound Instability

The return from prolonged fasting is often imagined as the corrective phase of the event, as though instability belongs only to deprivation and stability belongs automatically to restoration. That assumption is structurally inaccurate. Recovery is not inherently stabilizing. It is a transition, and transitions can destabilize when the system components involved do not realign at the same rate or within the same functional range. A prolonged fast does not end by handing the organism back to baseline. It hands the organism into a recovery process shaped by prior depletion, signaling distortion, downregulation, narrowed output, and accumulated recovery debt. That process can move toward stabilization, but it can also generate new forms of instability if its internal changes unfold out of sync with one another.

This is why recovery must first be understood as variable. There is no single "recovered state" that arrives uniformly across all domains once input returns. The system does not re-expand as one coordinated block. Different domains begin moving at different rates, with different sensitivities, and under different constraints inherited from the fasting state. Substrate handling may shift quickly in response to renewed availability, but signaling coherence may lag. Output may begin to rise before regulatory precision has been restored. Some adaptive suppressions may release faster than the structures that previously kept them in balance. Recovery therefore introduces variability rather than immediate stability because it involves multiple domains trying to re-enter broader function from a constrained baseline.

This variability matters because the organism is not restoring itself from a neutral pause. It is rebuilding coordination after a period in which that coordination was narrowed, distorted, or deferred. A system emerging from prolonged deprivation is not simply "starting up again." It is attempting to re-establish synchrony among processes that were altered by cumulative load and then challenged again during re-entry. That re-establishment does not occur evenly. Some systems become more responsive sooner. Others remain dampened or delayed. Some capacities expand before the structures that contain them have regained sufficient stability. Recovery therefore creates a state of transitional heterogeneity. The body is no longer fully organized around deprivation, but it is not yet broadly restored either.

This transitional heterogeneity is the foundation of rebound instability. When domains recover at different rates, mismatch emerges. Mismatch is not incidental. It is the mechanism through which recovery becomes volatile rather than stabilizing. The system is not destabilized simply because things are changing. It is destabilized because different parts of the system are changing on different timelines and with different magnitudes, causing their relationships to one another to become temporarily misaligned. A body in this state is not merely fluctuating. It is operating through uneven re-expansion, and that unevenness can distort regulation.

The clearest way to understand this distortion is through the concepts of overshoot and undershoot. Overshoot occurs when a recovering domain exceeds the range within which its return would remain stably integrated with the rest of the system. It is not merely "too much" in a simplistic sense. It is an expansion that outpaces coordination. A process becomes more active, more responsive, or more intense before the surrounding systems are ready to absorb that change cleanly. Undershoot is the opposite problem. It occurs when recovery remains incomplete, constrained, or delayed to a degree that prevents the system from re-establishing adequate function. In this case, the organism remains partially trapped in the logic of the fasting state even as other domains begin moving outward.

Both overshoot and undershoot create instability because both represent mismatch between what a domain is doing and what the broader system can currently support. Overshoot can generate volatility by pushing one layer of regulation beyond the stabilizing capacity of the others. Undershoot can generate instability by leaving one layer too narrow or under-responsive relative to the demands being placed on it by the rest of the recovering organism. In both cases, the problem is not that change is occurring. The problem is that change is occurring without sufficient alignment. Recovery is therefore not neutral. It can itself distort system behavior when the movement back toward broader function exceeds or fails to match current system capacity.

This is why rebound instability should not be interpreted as random fluctuation or as proof that recovery is somehow failing in a global sense. The system is still trying to restore itself. But it is doing so through a field of uneven expansion. One domain may be moving too quickly, another too slowly, and a third may be responding in a way that would be appropriate under baseline conditions but becomes destabilizing when paired with the current state of the others. The organism is not simply becoming "better" or "worse." It is becoming mismatched.

Once this mismatch is present, it begins to express itself in recognizable patterns. Instability after prolonged fasting does not emerge as a single generic disturbance. It appears through fluctuating coordination across signaling, output, and substrate handling. Signaling may become more reactive in one interval and more dampened in another because the hormonal environment is still recalibrating. Output may rise in one domain while remaining constrained in another, producing an uneven relationship between system demand and system support. Substrate handling may shift rapidly toward renewed intake while regulatory processes that govern allocation and stabilization remain partially organized around conservation. These fluctuations are not disconnected. They are different surfaces of the same underlying issue: domains are recovering out of sync.

This patterned quality is important because it shows that rebound instability is not mere noise. It reflects structure. A system in rebound volatility is not simply malfunctioning without logic. It is expressing the consequences of misaligned recovery. The organism is attempting to re-establish coherence, but because the elements required for coherence are not yet synchronized, the resulting behavior appears uneven. One of the defining features of this phase is inconsistency. The body does not respond with the same steadiness it would exhibit in a more stable state because the conditions required for that steadiness have not yet been fully rebuilt.

This inconsistency can be particularly misleading because some aspects of recovery may appear improved in isolation. A single domain may seem more active, more responsive, or more available than it was during the fast itself. But local improvement does not guarantee system stability. In fact, isolated improvement can become a source of volatility if it is not matched by the broader regulatory environment. This is one of the reasons rebound instability can be mistaken for positive adaptation when viewed too narrowly. A recovering system may show signs of increased capacity in one area while simultaneously becoming less coordinated overall. What matters is not whether a given domain has expanded. What matters is whether that expansion remains integrated within a recoverable system range.

This is where the oscillation envelope becomes essential again. The envelope does not apply only to fasting stress itself. It also governs recovery. A system stabilizes not merely by moving away from deprivation, but by re-expanding within the range where oscillation remains recoverable and coherent. Recovery becomes unstable when its movements fall outside that range. Overshoot pushes the system above the upper boundary of the envelope. This occurs when responsiveness, allocation, or output expand beyond what the current level of system coordination can support without volatility. Undershoot remains below the lower boundary. This occurs when recovery remains too narrow, too incomplete, or too suppressed to restore sufficient range for stable function.

In both cases, the problem is envelope violation. The system is no longer moving within a band of reorganization that can be absorbed and integrated cleanly. Instead, it is moving outside the recoverable range. Above the upper boundary, the system becomes over-reactive, over-amplified, or insufficiently contained. Below the lower boundary, it remains under-restored, under-responsive, or unable to regain enough range for coherent stabilization. Neither condition represents successful recovery, even though both may contain elements of change that look superficially like progress. Recovery stabilizes only when it remains inside the envelope—when the pace, magnitude, and coordination of re-expansion are matched to what the organism can currently integrate without distorting itself.

This is why recovery after prolonged fasting cannot be treated as inherently corrective. Recovery is a system event in its own right. It has dynamics, thresholds, mismatch potential, and volatility risks. The body is not guaranteed to move from depletion into stability just because input has returned and adaptation to absence has ended. It must move through a recovery process whose own alignment determines whether the system broadens coherently or destabilizes under the weight of uneven re-expansion. Recovery therefore belongs inside the same systems framework as fasting itself. It is not the simple opposite of stress. It is another phase in which system behavior is governed by timing, load, sensitivity, and coordination.

Seen clearly, rebound instability is not an exception to recovery. It is one of its predictable expressions when recovery is misaligned.

Variability in re-expansion creates mismatch. Overshoot and undershoot amplify that mismatch by pulling domains outside the range where their return can remain integrated. The resulting fluctuations across signaling, substrate handling, and output are not random disturbances but patterned consequences of a system trying to recover without yet being synchronized. And when that recovery moves outside the oscillation envelope, the body does not stabilize simply because deprivation has ended. It continues cycling through volatility until re-expansion returns to a range the system can actually contain.

Capacity Constraints

Prolonged fasting is often interpreted as though duration were limited primarily by willingness, discipline, or tolerance. That interpretation confuses subjective endurance with structural capacity. A biological system is not governed by how strongly it intends to continue under stress. It is governed by how much stress it can absorb while remaining sufficiently coherent to preserve regulation, adaptation, and recoverability. Capacity is therefore the correct starting point for understanding prolonged fasting. Not because it adds caution to the conversation, but because it defines the actual boundaries within which the event remains structurally containable. A system can continue only so long as the load imposed by duration remains inside what that system can absorb without losing coherence.

Capacity, in this context, should be understood as the total functional bandwidth available to the organism under current conditions. It is not simply energy reserve, nor is it reducible to a single physiological marker. It is the combined ability of the system to remain organized while under stress. A system with higher available capacity can absorb more disruption, sustain more transition, and preserve more coherence before destabilization begins to rise. A system with lower available capacity reaches its boundaries sooner. This is why capacity is the correct limiting variable. It determines not whether stress exists, but whether stress remains containable. The significance of prolonged fasting lies not only in the fact that it imposes stress, but in the fact that it imposes stress continuously enough to test the system's available bandwidth over time.

That bandwidth is finite. This point matters because prolonged fasting is often discussed as though stress could continue accumulating so long as the organism remains conscious, functional, or superficially responsive. Biological systems do not work that way. Every organism operates within a bounded range of recoverable strain. There is a limit to how much depletion, signaling distortion, downregulation, recovery compression, and cumulative burden can be integrated before coherence begins to narrow. Capacity is the name for that bounded range as it exists in the present state of the system. Once capacity is exceeded, the organism does not simply become "more stressed" in a generic sense. It begins losing the structural room required to absorb further stress without distortion.

This is why capacity must also be understood as state-dependent. It is not a fixed property that can be inferred once and then treated as stable across all conditions. Capacity changes with system condition because the organism's available bandwidth is shaped by what it is already carrying. A system with broad recovery margin, coherent signaling, low accumulated burden, and flexible substrate handling does not relate to fasting stress in the same way as a system already narrowed by unresolved strain, distorted regulation, or prior recovery debt. The same duration therefore does not mean the same thing in every system because the stress imposed by duration is being received by different amounts of available bandwidth. Capacity is not abstract potential. It is current absorbability.

This is where misunderstanding often enters. Stress is frequently interpreted as though it acts on an empty system, as though the body begins each event from a neutral baseline with full available room to absorb what follows. In reality, systems begin from where they are. Baseline condition is not background detail. It is part of the load-bearing architecture that determines whether duration remains tolerable or becomes destabilizing. A system already operating near its limits does not require dramatic additional burden to lose coherence. It only requires enough continued stress to exceed what little bandwidth remains. This is why prolonged fasting cannot be evaluated through duration alone. Duration is meaningful only relative to the capacity available to contain it.

Capacity is shaped from both within and without. Internally, one of the most important constraints is recovery capacity. Recovery determines how effectively the system can restore what stress consumes and how much burden can be absorbed without being carried forward as unresolved debt. A system with broad recovery capacity can absorb disruption with less residual strain because restoration remains available as an active stabilizing process. A system with narrow recovery capacity cannot clear the same burden as effectively. The same stress therefore occupies more of its total bandwidth because less of that bandwidth can be restored once consumed. Recovery capacity is not simply relevant after the fast ends. It is part of what determines how much of the fast can be structurally contained while it is happening.

Metabolic flexibility is another internal constraint because prolonged fasting requires the system to transition across substrate conditions without excessive friction. A system that can move more coherently through glycogen depletion, increasing fat reliance, and deeper internal resource dependence preserves more organizational continuity as the event unfolds. A system with lower flexibility must expend more adaptive effort simply to remain coherent across the same transitions. That additional effort is not neutral. It consumes capacity. This means that metabolic flexibility does not merely affect the "experience" of fasting. It changes the amount of available bandwidth the organism must spend in order to remain regulated while the event continues.

Signaling coherence is equally important because capacity depends not only on what resources are available, but on how accurately the system can coordinate itself under changing conditions. A body with clearer timing, more proportionate hormonal response, and more reliable internal communication can preserve coherence under stress with less distortion. A body with already narrowed or unstable signaling enters prolonged fasting with less coordination reserve available. This matters because signaling distortion does not only appear after capacity has been exceeded. It also contributes to how quickly capacity is consumed. A system that must regulate through imprecise or increasingly reactive signaling spends more bandwidth simply maintaining order.

These internal constraints do not operate in isolation. They are shaped continuously by external conditions that determine how much capacity is actually available in context. Sleep is one of the most important of these because it influences restoration, signaling precision, autonomic regulation, and the broader capacity of the system to recover from load. A system under prolonged fasting is not receiving stress only from the fast itself. It is receiving stress through the quality of the environment in which the fast is unfolding. If sleep is not supporting restoration, the organism does not merely feel more burdened. It actually has less available bandwidth with which to absorb continued deprivation.

Stress load from outside the fast functions similarly. The body does not separate the burden of fasting from the burden of everything else it is carrying. Physiological and environmental load are integrated through the same regulatory architecture. This means that external stress is not a parallel issue. It is part of the total system burden. A prolonged fast occurring within an already loaded context therefore consumes capacity faster than the same fast would inside a lower-load context, not because the fasting event has changed in abstraction, but because the system receiving it has less uncommitted bandwidth available to contain it.

Activity demands matter for the same reason. A system required to sustain higher output while under prolonged deprivation must allocate capacity across both the fasting event and the external requirements imposed upon it. The organism is therefore not only managing internal transitions. It is also preserving function under active demand. This increases total system burden and narrows the margin between current load and total available capacity. What matters is not whether the demand is "reasonable" in isolation. What matters is that capacity is shared across all demands. The same system bandwidth is being asked to support fasting, regulation, recovery limitation, and external output at once.

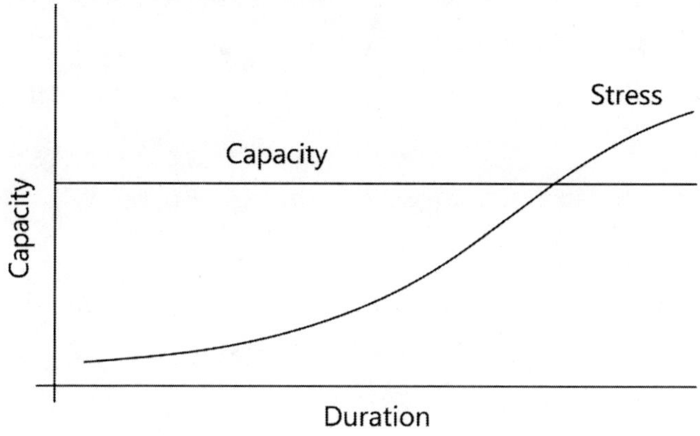

[Figure 11: Capacity vs stress crossover—rising stress curve intersecting fixed capacity threshold]

Once capacity is understood as finite, state-dependent, and context-shaped, the mechanism of instability becomes much clearer. Duration increases stress load, but capacity does not scale upward simply because the event continues. This asymmetry is central. Stress compounds over time through depletion, adaptive strain, signaling distortion, downregulation, re-entry sensitivity, and recovery debt. Each of these mechanisms adds burden or narrows the system's ability to contain burden. Capacity, by contrast, is not a variable that expands in parallel with prolonged exposure. In many cases, it narrows as duration continues because the event itself consumes the very bandwidth required to contain it. This is why duration cannot be understood as a neutral extension. It is not simply "more of the same." It is more load imposed onto a system whose available room to absorb that load may already be shrinking.

Mismatch emerges when the cumulative burden of the event exceeds the bandwidth the system still has available. This is not necessarily a dramatic threshold in appearance. It can begin as increasing friction, narrower tolerance, slower correction, more reactive signaling, or reduced ability to stabilize after small deviations. But all of these are expressions of the same structural problem: the system is carrying more load than it can absorb cleanly. Stress has overtaken capacity. The organism is no longer operating from within a recoverable range of strain. It is operating from within a mismatch between what is being imposed and what can still be integrated.

This mismatch is especially easy to miss because biological systems compensate. Compensation allows the organism to preserve outward continuity even after capacity has begun to be exceeded. It redistributes burden, narrows function, and preserves essential regulation long enough to delay obvious breakdown. But compensation does not erase mismatch. It conceals it temporarily by allowing the system to continue under conditions it can no longer absorb broadly. This is one of the reasons prolonged fasting can appear structurally manageable beyond the point at which the system has already begun losing coherence. Continuation is not proof of containment. A system can continue while already exceeding what it can stably support.

This is where breakdown risk becomes the correct frame. Breakdown should not be understood only as acute collapse or dramatic failure. In systems terms, breakdown begins earlier as progressive loss of coherence. The body remains active, but less broadly organized. Signaling becomes less precise. Output becomes more conditional. Recovery becomes less complete. Small deviations become more consequential. Thresholds are reached sooner. None of these changes need to present as catastrophic in order to represent the beginning of breakdown. They indicate that the system is no longer operating with enough bandwidth to preserve stable coordination under the load it is carrying.

This progressive quality matters because it explains why capacity exceedance is often misunderstood. If breakdown is imagined only as a sudden event, then everything short of that event appears tolerable by default. But the real issue is not whether collapse has occurred. It is whether the system is still operating inside a recoverable range of coherence. Once capacity has been exceeded, instability increases even if acute failure has not yet occurred. The organism becomes more vulnerable not because it has suddenly become weak, but because it has less remaining room to absorb any additional burden without destabilizing further.

Seen clearly, prolonged fasting is governed by capacity, not by intent, conceptual tolerance, or abstract resilience. Capacity defines how much cumulative stress the organism can absorb while preserving coherence. That capacity is finite, state-dependent, and shaped by both internal architecture and external context. Duration increases load, but capacity does not scale to meet duration simply because the event continues. When the burden of prolonged fasting exceeds the system's available bandwidth, mismatch emerges. Compensation may temporarily preserve continuity, but instability rises as coherence narrows. The limiting factor is therefore not how long the event can be conceptually sustained. It is how much the system can structurally contain before continued duration stops being absorbable and begins becoming destabilizing.

Aging Constraints

Aging is often described in metabolic writing through the language of decline, as though the central fact of time were that the system simply becomes weaker, slower, or less capable in some generalized sense. That framing is too blunt to be structurally useful. It collapses a systems problem into a mood. The more accurate interpretation is narrower and more specific. Over time, the organism tends to operate within a smaller range of recoverable variation. This is not best understood as deterioration in a dramatic or moralized sense. It is best understood through the Constraint Narrowing Model. The key shift is not that the body suddenly loses all capacity. It is that the bandwidth through which stress can be absorbed, organized, and restored becomes less expansive. In the context of prolonged fasting, this matters because the event is governed by how much strain the system can contain without losing coherence. If the system's bandwidth narrows, the same fasting stress does not carry the same meaning.

Capacity, in this framework, is not a trait in the abstract. It is the amount of organized flexibility available to the system at a given time. It includes the ability to tolerate depletion, maintain signaling coherence, preserve output within recoverable bounds, and restore stability after disruption. Earlier in life or under broader system conditions, that capacity may allow a wider range of oscillation before threshold pressure becomes destabilizing. Over time, however, that range often becomes smaller. This does not mean the organism stops adapting. It means adaptation occurs inside a narrower field. There is less room for deviation, less excess bandwidth for compensatory movement, and less available margin for absorbing stress without distortion. The significance of aging in prolonged fasting is therefore not that fasting becomes conceptually "harder" in some generic sense. It is that the same duration or same amplitude of deprivation now occupies a larger proportion of the system's total available range.

This narrowing becomes especially important because capacity is state-dependent rather than fixed. The body does not carry a static tolerance level forward unchanged across time. It carries a changing architecture of restoration, flexibility, and recoverability. Projecting past tolerance onto the current system therefore creates a structural error. A previous exposure that remained inside the system's bandwidth under earlier conditions does not automatically remain inside it later. The load imposed by prolonged fasting may be similar in external form, but its relative size changes when the system receiving it has less adaptive room. This is one of the central ideas of the Constraint Narrowing Model: stress does not need to become larger in order to become more destabilizing. It only needs to occupy a larger fraction of the system's current capacity.

Adaptive flexibility is one of the first aspects of this narrowing. A system with broader flexibility can move through changing substrate conditions, altered signaling environments, and variable recovery demands with less friction. It can absorb temporary mismatch without immediately converting that mismatch into instability. As capacity narrows over time, that flexibility often becomes more conditional. The organism can still adapt, but adaptation requires a tighter set of conditions and may produce less excess room once established. This means that prolonged fasting does not simply encounter a system with the same functional logic it always had. It encounters a system that may still be coherent, but coherent within a more bounded range. That boundedness changes how stress is distributed, how much variation can be tolerated, and how easily the organism can return to a stable center after being displaced.

Tolerance range narrows for the same reason. Tolerance is not a matter of determination. It is the width of the range inside which the system can remain organized under stress without entering destabilizing compensation. When that range becomes smaller, the same fasting event presses more quickly against meaningful boundaries. A duration that once remained inside a broad absorbable zone may now approach threshold sooner because the system's available room has contracted around it. This is why aging must be framed as structural narrowing rather than decline narrative. The issue is not that the organism has become defective. The issue is that the geometry of tolerance has changed. The same exposure now occupies more of the available field.

This becomes even more consequential once recovery is considered, because recovery is one of the primary mechanisms through which capacity is restored after stress. A system does not preserve bandwidth simply by enduring perturbation. It preserves bandwidth by being able to restore what perturbation consumes. Recovery is therefore not secondary to capacity. It is part of what sustains capacity across time. When recovery slows or becomes less complete, the organism loses one of its most important methods for re-expanding after stress. This means that capacity narrowing is not only about what happens during the fast itself. It is also about what happens afterward, when the system attempts to restore what the event has compressed.

Recovery latency is one of the clearest changes here. Latency refers to the time required for the organism to move from post-stress reactivity back toward a more coherent baseline. In a system with broader restoration bandwidth, this transition can occur with relatively more efficiency because signaling, substrate handling, and output are able to re-synchronize with less residual distortion. As recovery latency increases, the system remains in a post-stress transitional state for longer. This does not simply extend discomfort or prolong visible effects. It means the organism remains structurally narrowed for a longer period after the event has ended. The low-input configuration does not fully unwind at the same rate, and the system therefore carries the consequences of the fast forward more persistently.

Recovery completeness is equally important. Restoration is not only a matter of whether the organism begins moving back toward baseline. It is a matter of how fully it arrives there. A system can partially recover while still carrying unresolved burden in signaling coherence, substrate regulation, output range, or adaptive flexibility. When recovery becomes less complete, the organism does not re-enter future stress from a fully restored baseline. It begins from a reduced one. This is where the relationship to recovery debt becomes structurally important. Recovery debt is not simply a result of repeated fasting events in the abstract. It is also shaped by the degree to which restoration after each event remains partial. If recovery depth decreases over time, then the gap between what depletion requires and what restoration delivers becomes easier to accumulate.

Reduced restoration efficiency therefore matters because it changes how much capacity is truly available for subsequent perturbation. A system that restores less completely is not only slower to recover. It is narrower when it does recover. The bandwidth that once returned more fully after stress now returns only partially, leaving behind a more constrained baseline. This means that prolonged fasting does not need to become more severe externally in order to become more consequential internally. The event is being received by a system whose restorative return has become shallower. The result is that cumulative burden is more likely to persist, and the organism becomes increasingly shaped by what prior stress has not fully released.

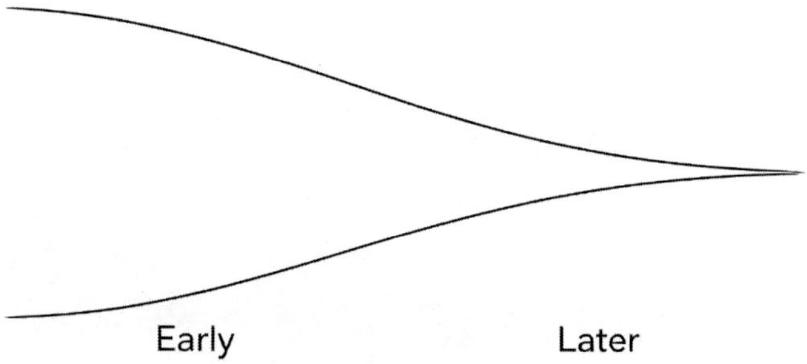

[Figure 12: Envelope contraction—wide oscillation range narrowing over time]

This is the point at which the oscillation envelope becomes indispensable again. The oscillation envelope was introduced earlier as the recoverable range within which stress and recovery can occur without driving the system into lasting instability. It is the zone in which variation remains absorbable and re-stabilization remains possible. Under the Constraint Narrowing Model, the significance of aging is that this envelope contracts over time. The organism does not stop oscillating between stress and restoration, but the width of the band inside which those oscillations remain recoverable becomes smaller. This means there is less room for amplitude, less room for mismatch, and less room for recovery variation without threshold pressure becoming more relevant.

Envelope width is one of the most useful ways to conceptualize tolerance because it reveals that the problem is not always the stressor itself. A stress amplitude can remain externally unchanged while becoming internally more destabilizing if the envelope around it has narrowed. This is one of the most important clarifications in the chapter. Risk increases not necessarily because fasting stress becomes objectively larger, but because the system's tolerance field around that stress becomes smaller. The same event is now operating inside a reduced recoverable range. What once fit comfortably within the envelope may now occupy a much larger proportion of it, bringing the system closer to instability without any visible change in the stressor's external form.

This reduced room for deviation changes the meaning of variation. In a wider envelope, the system can tolerate temporary mismatch, uneven transitions, or partial distortion without immediately approaching threshold. In a narrower envelope, those same deviations become more consequential because they occupy more of the available range. The organism therefore becomes less forgiving of inconsistency, not because it has become fragile in a generalized sense, but because the distance between ordinary variation and destabilizing strain has become shorter. The envelope has less excess room to absorb what once remained easily recoverable.

This is also why the same stress exposure becomes more loaded over time. A prolonged fast that once represented a moderate perturbation may later represent a far more significant percentage of the system's total available tolerance. Relative load increases even if absolute load does not. That distinction matters because systems are governed by relative burden, not by abstract external comparisons. The organism does not respond to duration in a vacuum. It responds to how much of its current bandwidth that duration consumes. Once the envelope narrows, the same fasting event takes up more of the available field, leaving less room for recovery error, less room for compensatory movement, and less room for maintaining coherence if the event deepens or if re-entry becomes unstable.

From here, increased instability risk becomes easier to understand. Thresholds are reached earlier because the distance between baseline and destabilization has shortened. The system does not need to be pushed as far to encounter meaningful instability because its recoverable range is smaller from the outset. This does not mean threshold crossing becomes inevitable. It means threshold pressure becomes more sensitive to the same exposure. A smaller displacement can now produce a larger relative consequence because the organism has less available room in which to absorb it.

This increased sensitivity changes how stress is distributed. The same fasting duration, the same depletion depth, or the same refeeding perturbation can now produce more reactivity, more mismatch, or more prolonged instability because the system has less excess bandwidth with which to contain the event cleanly. Destabilization therefore occurs faster not necessarily because the system has become chaotic, but because the path from perturbation to threshold has become shorter. There is less distance for the organism to travel before compensation begins narrowing function, less room for partial mismatch to remain harmless, and less recovery depth available to unwind what the stress has compressed.

The result is that prolonged fasting becomes a higher relative load over time even when it remains externally unchanged. This is one of the most important system clarifications in the chapter. It corrects the common assumption that prior tolerance can be projected indefinitely forward. Under the Constraint Narrowing Model, that projection fails because capacity, recovery efficiency, and envelope width do not remain static. The body is not receiving the same event with the same structural room to absorb it. It is receiving it through a narrower bandwidth, a slower restoration process, and a smaller oscillation field.

Seen clearly, aging is best understood here not as a decline story, but as progressive narrowing of recoverable system range. Capacity becomes more bounded, recovery becomes slower and less complete, and the oscillation envelope contracts around the same external stress. The result is not that the organism suddenly becomes incapable. It is that the same fasting exposure now carries more relative load and approaches instability sooner because the room for containing it has become smaller. That is the core of the Constraint Narrowing Model: tolerance decreases over time not because stress necessarily intensifies, but because the system's ability to absorb, restore, and remain coherent within that stress becomes increasingly narrow.

Error Margins & Compensation

Stability is often imagined as the absence of error, as though a biological system remains coherent only when deviation is minimal and all variables stay tightly aligned. That interpretation misrepresents how living systems actually preserve order. Biological stability does not depend on perfection. It depends on margin. Margin is the recoverable operating space within which variation, mismatch, delay, or temporary distortion can occur without pushing the system into instability. It is the buffer that allows the organism to remain coherent while not being exact. A system with sufficient margin can absorb irregularity, tolerate imperfect transitions, and recover from partial deviation without immediate threshold pressure. A system with little margin cannot. The same deviation that would once have remained structurally trivial can become destabilizing once the available room around it has narrowed.

This is why margin must be treated as a core systems property rather than as a secondary consequence of other mechanisms. A prolonged fast does not challenge the body only through depletion depth, signaling change, or recovery demand. It also challenges the amount of recoverable room the system still has while those changes unfold. Margin is what determines whether those changes remain absorbable. It is the space within which the organism can remain slightly misaligned and still return to coherence without escalating into more serious instability. The significance of prolonged fasting is that it progressively consumes this space. Stress does not only accumulate as burden. It accumulates as a reduction in the system's tolerance for error.

This makes margin closely related to the oscillation envelope introduced earlier, though the two are not identical in emphasis. The envelope describes the broader recoverable range within which stress and recovery can cycle without destabilizing the system. Margin refers more specifically to the local buffering space inside that range—the amount of deviation the system can tolerate before a fluctuation, mismatch, or disturbance becomes meaningful. The distinction matters because a system may still be technically inside its broader envelope while already operating with very little local margin. In such a state, it has not yet fully destabilized, but it has become much more sensitive to ordinary variation. This is one of the most important features of prolonged fasting as a time-dependent event: not only can the system be pushed closer to threshold, it can also lose the buffering space that once prevented small deviations from mattering.

The Margin Compression Model becomes useful precisely here. The central claim of the model is not that all systems fail because of extreme stress alone. It is that tolerance for error decreases as cumulative load and constraint narrow the system's recoverable range. In other words, instability often emerges not only because stress becomes large, but because the space available for containing ordinary imperfection becomes small. This is what makes prolonged fasting structurally different from a short perturbation. A brief event may challenge the system while leaving enough margin intact for minor mismatch to remain harmless. A prolonged event can compress that margin until even small deviations begin to consume a meaningful portion of what remains.

Cumulative load is one of the primary drivers of this compression. As stress compounds through depletion, adaptive strain, downregulation, signaling distortion, and recovery debt, the organism does not merely become "more burdened." It becomes less buffered. Each unresolved layer of strain narrows the room the system has to absorb the next one. A system carrying substantial cumulative load is therefore not only closer to threshold in a broad sense. It is also operating with less slack. The distance between minor variation and meaningful consequence becomes shorter. This is why cumulative load matters beyond simple burden accounting. It changes the geometry of stability itself.

Depletion contributes to this process because it reduces the energetic and regulatory flexibility available to contain mismatch. A system with more accessible reserve and more flexible substrate handling can tolerate imperfect transitions with less friction. A depleted system must operate more tightly because it has less room to redistribute load without cost. The issue is not merely that less energy is available. It is that the system has less excess operating room with which to absorb the consequences of imperfection. Once depletion deepens, the organism must maintain coherence with fewer resources and less flexibility, which means the same deviation now consumes a larger proportion of its remaining recoverable space.

Signaling distortion reinforces the same compression from another angle. A body with clearer timing, more proportionate hormonal movement, and more reliable internal communication can absorb small mismatches more quietly because its correction mechanisms remain more precise. Once signaling becomes narrowed, exaggerated, delayed, or otherwise distorted, the organism loses some of that precision. It must regulate through noisier or less proportionate inputs. This means that small deviations are less likely to be corrected cleanly and more likely to propagate. Compression here does not require catastrophic signaling failure. It only requires enough loss of clarity that ordinary variation begins to have larger downstream consequences because the system has less precise means of containing it.

Recovery limitation adds another layer because margin is not preserved by stress tolerance alone. It is also preserved by how fully the system can restore itself after stress. A body that recovers more completely can re-expand its available operating space after perturbation. A body that recovers only partially remains narrowed. This means that margin compression is not just an in-event phenomenon. It can persist and accumulate when restoration fails to fully reopen the room that stress has consumed. Recovery debt is therefore one of the ways compressed margin becomes chronic rather than transient. The system does not simply become sensitive during the fast. It may remain more sensitive afterward if restoration does not fully rebuild what the event has narrowed.

The key clarification here is that compression does not mean immediate failure. A system can continue functioning while its margin is already significantly reduced. This is one of the reasons margin compression is so easy to overlook. The organism may still appear coherent, still perform within a recognizable range, and still preserve outward continuity. But beneath that continuity, the amount of error it can absorb without destabilizing may already be much smaller than before. The body has not collapsed. It has become less forgiving. That change is often invisible until ordinary variation begins producing effects that appear disproportionate relative to their size. The Margin Compression Model explains why those effects are not actually disproportionate at the systems level. They are proportionate to the amount of margin that remains.

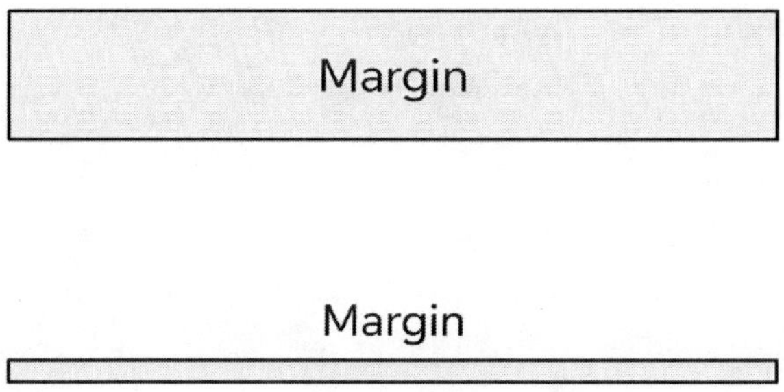

[Figure 13: Wide vs narrow margin—broad stable range compressing into narrow tolerance band]

Once margin has compressed, sensitivity to deviation rises. This is one of the most important transitions in the chapter because it explains why the same system can behave very differently under what appears to be the same level of disturbance. A system with broad margin can absorb a mismatch quietly because the deviation occupies only a small fraction of the available tolerance band. A system with narrow margin cannot do the same because the deviation now takes up a much larger proportion of the remaining buffer. The disturbance itself has not necessarily changed. What has changed is the relationship between the size of the deviation and the size of the space available to contain it.

This is why small deviations can begin producing larger effects. The effect is not large because the deviation is inherently extreme. It is large because the system no longer has enough room to keep that deviation small in consequence. A modest inconsistency in substrate handling, a temporary signaling mismatch, a slight disruption in recovery sequencing, or a small perturbation in output may all remain relatively inconsequential in a broad-margin system. In a compressed system, however, those same irregularities become more structurally meaningful because there is less available room to buffer them. Instability therefore emerges not from the absolute size of the error, but from its relative size compared to the margin that remains.

This relative relationship is essential. It corrects one of the most common misunderstandings in metabolic stress analysis, which is the assumption that only large disturbances matter. In reality, the importance of a disturbance depends on context. A small error inside a broad tolerance band remains small. The same error inside a narrow band can become threshold-relevant. This is why systems under prolonged fasting can begin exhibiting high-sensitivity behavior without any dramatic new insult. The organism is not necessarily being hit with more extreme inputs. It is being asked to manage ordinary variation with less recoverable room.

As this happens, the system becomes more reactive. Reactivity here does not simply mean that signals become stronger or responses become more dramatic. It means that the organism has less tolerance for mismatch before meaningful adjustment is required. The body begins operating in a tighter, less forgiving mode because more of its stability depends on minimizing deviation it no longer has enough room to absorb. This is one of the structural reasons instability can appear to accelerate late in a prolonged fast or under repeated exposure. The system is not merely more stressed. It is more tightly constrained around its remaining margin.

This tightening also changes the behavior of thresholds. When margin is broad, the organism can move through fluctuation without quickly approaching threshold because there is enough buffer between ordinary variation and destabilizing transition. When margin is compressed, that buffer becomes thinner. Threshold crossing therefore occurs faster because less displacement is required. The system does not need a dramatic push to become unstable. It only needs a disturbance large enough relative to the narrow margin that remains. This is why threshold behavior often appears more abrupt in constrained systems. The transition feels sudden not necessarily because the disturbance was extreme, but because the organism had very little room left before the disturbance became consequential.

Reduced tolerance for mismatch is one of the clearest outcomes of this process. A system with compressed margin cannot accommodate uneven recovery, imperfect sequencing, delayed correction, or partial distortion as quietly as a broader system can. The same forms of inconsistency that would once have remained well inside recoverable range now begin pressing more quickly against meaningful boundaries. This is one of the reasons prolonged fasting can create a state in which ordinary system noise becomes functionally important. The body is not being destabilized by absurdly large errors. It is being destabilized by the loss of the buffer that once kept ordinary error from mattering.

Disproportionate system responses follow from the same logic. Once margin is small, responses can appear excessive relative to the trigger that preceded them. But what appears excessive from the outside is often proportionate to the internal state of the system. A compressed organism must respond more aggressively or more visibly to smaller disturbances because it has less room for passive absorption. The response is not irrational. It is the behavior of a system operating near the edge of its remaining tolerance band. This is one of the most important explanatory powers of the Margin Compression Model: it shows why instability can emerge without extreme inputs and why "small things" begin to matter once the recoverable range has narrowed enough.

Seen clearly, stability depends less on eliminating all deviation than on preserving enough margin to contain deviation without escalation. Prolonged fasting challenges this condition by progressively compressing the recoverable room around the system. Cumulative load, depletion, signaling distortion, and recovery limitation all contribute to the loss of buffering capacity, not necessarily by causing immediate failure, but by narrowing the amount of mismatch the organism can tolerate before instability begins to rise. Once that margin becomes small, sensitivity increases. The same deviations produce larger effects because they now occupy a larger fraction of the remaining tolerance band. Thresholds are crossed faster, mismatch becomes harder to absorb, and the system begins responding more dramatically not because the inputs have become extreme, but because the buffer around them has been lost. That is the central logic of the Margin Compression Model: instability emerges not only from magnitude, but from the shrinking space within which magnitude can still be safely contained.

Failure Cascades

Failure is often imagined as an endpoint, a singular event in which the system crosses from functional to nonfunctional in one decisive movement. That interpretation is too coarse to explain how instability actually behaves in interconnected physiology. Systems rarely move from coherence to collapse in one isolated jump. More often, they destabilize through sequence. A contained disturbance ceases to remain contained, begins altering adjacent conditions, and then spreads through connected pathways until multiple domains are no longer able to regulate in coordinated relation to one another. This is the basis of the Failure Cascade Mechanism. More specifically, it can be described through the Cascade Failure Model: a localized instability propagates through interconnected systems, transforming a contained disruption into multi-domain instability. The importance of this model is that it shifts failure from the language of isolated breakdown to the language of transmission. The question is no longer only what failed. The question becomes how failure moved.

This matters because prolonged fasting is not a single-variable event. It is an accumulating, state-dependent stress condition unfolding inside an organism whose regulatory domains are deeply interdependent. Substrate handling, hormonal signaling, output regulation, recovery processes, autonomic tone, and adaptive compensation do not operate as independent modules that can fail cleanly without consequence to one another. They are linked. A disturbance in one domain alters the operating conditions of the others. This means that instability does not need to begin everywhere in order to become system-wide. It only needs to begin somewhere that matters enough within a sufficiently constrained state. Once that occurs, the potential for cascade emerges.

The trigger point is where this process begins. A trigger should not be understood as the total cause of collapse, nor as the sole origin of the instability that follows. It is the first visible breach through which a broader destabilization process becomes active. In many cases, the trigger is localized. It may begin as a threshold crossing in signaling coherence, a mismatch in re-entry conditions, a failure of compensation, a disproportionate response to a relatively small perturbation, or a shift in substrate or output regulation that the system can no longer absorb cleanly. What makes the trigger important is not necessarily its size in isolation. It is the state of the system into which it occurs.

This distinction is critical because the same local event can remain trivial in one context and become system-relevant in another. A small disturbance inside a broad-margin, low-load system may be absorbed without visible consequence. The same disturbance inside a narrowed, debt-carrying, high-sensitivity system can become the opening through which instability begins to spread. This is why trigger magnitude often appears smaller than the outcome it eventually precedes. The outcome is not large because the trigger was inherently extreme. The outcome is large because the system had already lost enough buffering capacity that the trigger now occupied a meaningful proportion of what remained. Reduced margin, cumulative constraint, prior load, recovery limitation, and signaling distortion all increase the likelihood that a local disruption will become the first meaningful breach rather than a harmless fluctuation.

This is one of the reasons early instability is so often misread. Small disturbances are frequently interpreted as noise when they occur in isolation from obvious collapse. But within a constrained system, they may represent the first point at which the organism is no longer able to localize strain. A trigger event matters when it reveals that containment has weakened. The body is not failing because the disturbance exists. It is failing because the disturbance is no longer staying where it began. This is the first shift that the Cascade Failure Model requires the reader to understand. Localized instability is not important only for what it disrupts directly. It is important for what it changes next.

Once the trigger has breached local containment, propagation becomes the central issue. Propagation refers to the way instability spreads from one domain into adjacent ones through the system's existing interdependencies. This does not require mechanical transfer in a simplistic sense. It occurs because no major physiological domain operates without affecting the conditions under which the others must regulate. Signaling, resource allocation, and regulatory dependency form the primary pathways of this spread. A disturbance in one layer changes the state of the system in which neighboring layers are trying to maintain coherence. That change becomes the vehicle of propagation.

Signaling is one of the clearest propagation paths because hormonal and neural coordination do not merely reflect state. They help create it. When signaling becomes distorted in one domain, the rest of the organism begins receiving altered information about priority, timing, and response. A local disruption therefore becomes an altered regulatory environment for everything downstream of it. The organism is not only managing the original disturbance. It is now attempting to coordinate through a communication layer that has already been shifted by that disturbance. This means that the first failure begins changing the interpretive conditions for other systems before those systems have independently "failed" in any primary sense.

Resource allocation functions similarly. The body cannot contain instability without spending something. When one domain begins destabilizing, resources are often redirected toward preserving continuity there. But redirection is not free. What is allocated toward one layer is no longer equally available elsewhere. This means that the first instability does not remain metabolically local even if it appears so initially. It changes the economy of the system. Adjacent domains must now regulate under slightly altered energetic and functional conditions because the organism has already begun spending capacity to contain the first disturbance. Propagation therefore occurs not only because the original failure spreads directly, but because the system surrounding it is forced to reorganize around its existence.

Regulatory dependencies complete the picture. Many physiological functions are stable only because other functions remain sufficiently coherent to support them. When one of those supporting layers becomes unstable, the systems depending on it do not remain unaffected. Their own ability to regulate becomes conditionally compromised. This is one of the most important features of cascade behavior. Failure spreads not only through direct disturbance, but through dependency withdrawal. A domain does not need to be hit directly to become destabilized. It only needs to lose enough of the conditions it depends on in order to stop regulating cleanly.

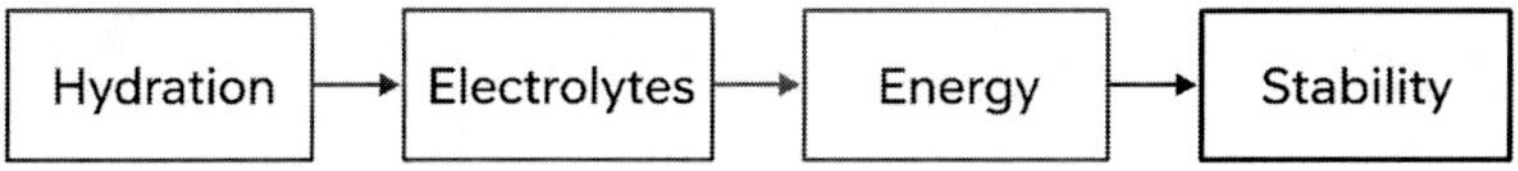

[Figure 14: Cascade chain]

This is what makes propagation directional but also reinforcing. It often begins in one place and moves outward through dependency pathways, but once multiple domains are involved, those domains begin feeding instability back into one another. A signaling disturbance alters substrate allocation. Altered substrate handling increases regulatory strain. Regulatory strain further distorts signaling. The system is no longer dealing with a single disturbance traveling through passive channels. It is now dealing with active interaction between partially destabilized systems that are each making the others harder to stabilize. Propagation therefore becomes recursive. The original trigger matters less over time because the instability is now being maintained and amplified by the interactions of the domains it has already touched.

At this stage, the cascade moves from local spread into multi-system impact. This is the point at which failures stop being meaningfully independent. Multiple systems are now affected at once, and the significance of each individual disturbance begins changing because it is no longer operating alone. A signaling issue is no longer merely a signaling issue if it is unfolding inside altered resource distribution, reduced output tolerance, compressed recovery bandwidth, and growing compensatory strain. Each domain now carries not only its own burden, but the burden imposed by its overlap with the others. This is where instability compounds.

Compounding is different from simple accumulation. Accumulation means multiple burdens are present. Compounding means those burdens begin interacting in ways that amplify one another. In a cascade state, overlap increases the difficulty of stabilization because no domain can be restored in isolation without being shaped by the instability still present in the others. This is why coordination begins degrading rapidly once multi-system involvement is established. The organism is no longer regulating through separable problems. It is regulating through a network of interacting distortions whose combined effect is larger than the sum of each one taken alone.

This interaction is one of the reasons cascades are often underestimated early and overestimated late. Early in the process, the trigger appears too small to explain the seriousness of the eventual outcome. Late in the process, the outcome appears too large to trace back clearly to the original trigger. Both impressions are products of the same structural misunderstanding: the assumption that systems fail only through isolated cause-effect chains. In reality, the trigger initiates a process whose later behavior is governed increasingly by interaction. Once multiple domains are destabilized, the system is no longer responding mainly to the first breach. It is responding to the interlocking burden of the failures now overlapping inside it.

Coordination degrades because coordination depends on the organism being able to sequence, prioritize, and synchronize across systems whose behavior remains at least partially interpretable. Once multiple layers become unstable together, the body loses not only resources and regulation but relational clarity. It becomes harder to determine what should be corrected first, what can still be buffered, and what response in one domain will produce destabilizing consequences in another. This is why cascades often feel nonlinear in their later phases. The organism is not simply becoming "more unstable" in a linear fashion. It is becoming less able to coordinate a coherent response to instability because the instability has become distributed and mutually reinforcing.

From here, collapse can be defined more precisely. Collapse is not merely the presence of many failures at once. It is the loss of coordinated system function. More specifically, collapse occurs when the organism can no longer stabilize itself within its recoverable range. The body may remain active, responsive, or even partially compensatory, but if its domains are no longer regulating in coherent relation to one another, the system has moved into collapse in a structural sense. This is one of the most important distinctions in the chapter. Collapse does not require total cessation. A system can still be visibly operating while no longer coherently organized.

This is why collapse should be understood as progressive rather than instantaneous. A cascade rarely produces one clean moment at which the system simply "fails" in an absolute sense. Instead, the organism loses coherence in layers. First, one disturbance stops remaining local. Then propagation begins. Then multi-system interaction develops. Then coordination degrades to the point that stabilization becomes increasingly difficult or impossible within the system's current range. By the time collapse is visible in a broad sense, it has usually been forming for some time through the accumulation and interaction of smaller destabilizations that were no longer being contained.

This progressive nature also explains why collapse is often missed until relatively late. Systems are built to compensate, and compensation can preserve visible function deep into instability. But compensation is not the same as coherence. It can mask mismatch while the underlying cascade continues spreading. The organism may still appear active because enough local functions remain online, yet the overall coordination required for stable self-regulation may already be failing. This is one of the defining dangers of cascade behavior. It creates the illusion that the system is still broadly intact while the architecture of containment is already eroding beneath that appearance.

Seen clearly, the Failure Cascade Mechanism closes an important conceptual gap in the systems analysis of prolonged fasting. It explains why instability cannot be understood solely through isolated thresholds, individual stressors, or local dysfunction. In a constrained organism, small disturbances can become meaningful because reduced margin and prior load make them harder to contain. Once containment is breached, instability spreads through signaling, resource allocation, and regulatory dependency pathways. That spread becomes self-reinforcing as multiple systems become involved, and interaction then amplifies the burden beyond what any one failure could explain in isolation. Collapse, in this model, is not a sudden event imposed from outside the system. It is the internal consequence of a contained disruption becoming uncontained, then propagated, then compounded until coordinated stabilization is no longer possible within the organism's remaining recoverable range.

Oscillation Envelope (CORE)

A biological system does not remain stable by staying still. It remains stable by moving in ways it can recover from. This distinction is the foundation of the Oscillation Envelope Mechanism and the reason the Oscillation Envelope Framework must sit at the center of any serious systems analysis of prolonged fasting. Physiology is not a fixed-point phenomenon. The organism does not operate by holding every variable at one static ideal and then losing stability whenever displacement occurs. It operates through continuous movement across changing conditions, changing loads, changing signals, and changing recovery demands. The question is therefore not whether stress occurs. Stress is part of ordinary system behavior. The question is whether the movement created by stress remains recoverable. That is what the envelope defines.

The oscillation envelope can be described as the bounded range within which stress and recovery can occur while preserving system coherence. It is not a metaphor for comfort, nor a simplified version of "balance" in a generic sense. It is a structural range. Inside that range, the organism can be displaced, can adapt, can deviate, and can still return with enough integrity intact that stability is preserved across time. Outside that range, movement ceases to be recoverable in a meaningful systems sense. The organism may still be active, compensatory, or superficially functional, but the coherence that allows repeated stress and recovery to remain sustainable begins to break down. This is why the framework matters. It allows stress and instability to be separated conceptually. Stress is not automatically destabilizing. Instability begins when stress and recovery no longer remain contained within a recoverable operating range.

This distinction is especially important in prolonged fasting because fasting is often interpreted through extremes. It is either imagined as an inherently beneficial stressor or as an inherently dangerous one, depending on which assumptions are carried into the discussion. Both framings miss the structural point. Stress is not meaningful by itself. Its meaning depends on where it occurs relative to the envelope. A stressor can be significant without being destabilizing if the system can absorb the displacement and return. Conversely, a stressor can appear relatively modest in external terms while still becoming destabilizing if it pushes the organism outside what it can recover from. This is why the envelope is more useful than absolute intensity. It organizes the system around recoverability rather than around abstract magnitude alone.

Oscillation, then, is not a sign of instability. It is a normal feature of living regulation. The system moves between load and restoration continuously. Energy availability changes. Signaling shifts. Output rises and falls. Recovery processes deepen and recede. This movement is not a failure of equilibrium. It is the form equilibrium takes in a living system. What matters is not whether oscillation occurs, but whether the oscillation remains contained. A stable organism is not one that avoids displacement. It is one that can move through displacement and return without accumulating distortion faster than it can resolve it. This is the central premise of the Oscillation Envelope Framework: stability is contained oscillation, not the absence of stress.

To understand how that containment works, the envelope must be treated as having two necessary boundaries. Stability does not depend on one limit alone. It depends on the relationship between stress amplitude and recovery depth. These are represented by the upper and lower bounds of the envelope. The upper bound defines the maximum stress amplitude the system can absorb without destabilizing. The lower bound defines the minimum recovery depth required to restore coherence after displacement. These are not mirror images of one another, and they do not describe the same function from opposite directions. They are distinct but interdependent requirements for stability. The system remains coherent only when both are respected.

The upper bound is the stress limit. It refers to how far the system can be displaced and still remain within a recoverable range. Stress amplitude can be understood as the size of the excursion created by a perturbation. A prolonged fast increases this amplitude over time through depletion, signaling alteration, downregulation, cumulative load, and capacity pressure. But the significance of that amplitude is not determined in abstraction. It is determined by whether the excursion remains inside what the organism can absorb without entering unrecoverable strain. Once stress amplitude exceeds the upper bound, the organism is no longer simply "under stress." It has moved beyond the range within which the system can remain broadly coherent under that load.

This is why the upper bound should not be confused with visible collapse. A system can exceed its upper bound before it appears acutely nonfunctional. Biological systems compensate. They preserve visible continuity deep into strain. But compensation is not the same as contained oscillation. Once the upper bound has been crossed, the organism may still be active, but it is no longer moving inside a range that allows stable return. The excursion has become too large relative to available capacity, and the body is now managing instability rather than recoverable stress. This is where overshoot begins, and overshoot matters because it shifts the system from displacement into destabilization.

The lower bound is different. It does not define how much stress can be tolerated. It defines how much return is required. Recovery depth refers to the extent to which the system must reconstitute coherence after displacement in order for the next cycle not to begin from accumulated distortion. A system does not remain stable simply because stress remains moderate. It also requires enough restoration to restore bandwidth, re-synchronize signaling, reopen margin, and reduce residual load. If that return remains too shallow, the organism may remain technically below the upper bound on any given excursion while still drifting toward instability over time. This is why the lower bound is indispensable. Stability requires not only survivable displacement, but sufficient return.

The lower bound therefore acts as the recovery requirement. It defines the minimum depth of restoration needed for the system to remain within a recoverable operating range across repeated cycles. Once recovery falls below this level, the organism begins carrying unresolved burden forward. Recovery debt forms. Margin compresses. Capacity narrows. The system may continue oscillating, but it is no longer oscillating cleanly. Each cycle begins from a more constrained state than the one before it. This is not acute failure, but it is a departure from stable containment. Undershoot begins here, and undershoot matters because it reveals that instability can emerge not only from too much stress, but from too little restoration.

This is one of the most important clarifications in the framework: both bounds are required for stability. A system can remain below the upper bound and still become unstable if it repeatedly fails to return above the lower bound of adequate recovery. Likewise, a system can recover deeply after a given stressor and still destabilize if the stress amplitude itself repeatedly exceeds what the organism can contain. Stability therefore cannot be reduced to "stress management" alone, nor to "recovery quality" alone. It is the relationship between displacement and return, bounded by both the upper and lower limits of recoverable oscillation.

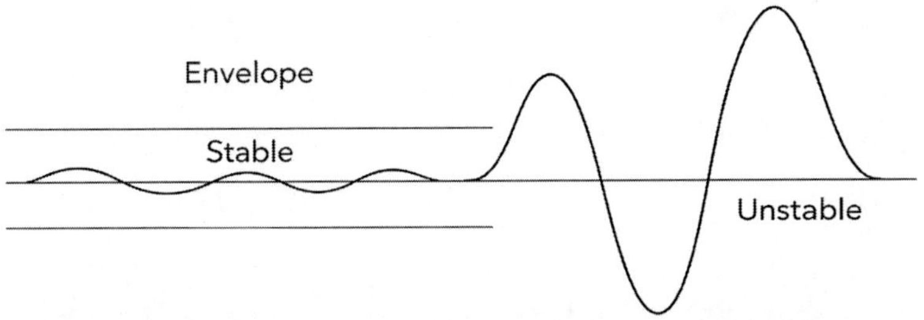

[Figure 15: Oscillation Envelope]

Once the envelope is defined this way, stable oscillation becomes easier to understand with precision. Stability is not stasis. It is repeated movement within bounds. The system experiences stress, is displaced by that stress, and then returns with enough coherence restored that the next displacement does not begin from progressively worsening distortion. This is the ordinary form of stable physiology. The body does not remain healthy because it avoids all perturbation. It remains coherent because perturbation and restoration remain linked inside a recoverable range.

This means that the organism does not need perfect alignment to remain stable. Perfection is not the requirement. Sufficient containment is. A system can tolerate variation, asymmetry, minor mismatch, and imperfect transitions so long as the total amplitude of displacement remains inside the upper bound and the depth of return remains above the lower bound. This is one of the reasons the envelope is such a useful model. It replaces the false expectation of precision with a more accurate expectation of recoverability. Stability does not require the body to be exact. It requires the body to remain within a range where exactness is unnecessary because ordinary deviation is still absorbable.

Stress creates displacement inside this framework. Recovery creates return. The organism remains coherent not because displacement never occurs, but because return remains sufficient to restore the conditions required for the next displacement to remain absorbable as well. This is what recoverable cycling means. The body moves away from center and then back toward it without losing too much of the structural room required to keep doing so again. Once that room begins shrinking faster than it is restored, stable oscillation begins giving way to unstable oscillation. The significance of prolonged fasting is that it places progressive pressure on both sides of this equation: it increases displacement amplitude while also often reducing recovery depth.

This dual pressure is what makes the envelope more powerful than any one mechanism discussed earlier. The organism is not destabilized by one factor alone. It is destabilized when multiple mechanisms converge on the same bounded structure from different directions. Depletion pushes upward on stress amplitude. Cumulative load increases displacement cost. Signaling distortion makes return less precise. Recovery debt reduces restoration depth. Margin compression narrows the room for ordinary error. Capacity narrowing lowers the total range itself. Each of these mechanisms is important, but none reaches its full meaning until it is understood relative to the envelope. The envelope is what gives them common geometry.

From here, instability can be defined with equal clarity. Instability is not simply the presence of stress. It is movement outside the envelope. More specifically, instability begins when the system either rises above the upper bound or fails to recover above the lower bound. These are the two primary failure directions, and each produces instability through a different pathway. Overshoot occurs when stress amplitude exceeds capacity. Undershoot occurs when recovery remains too shallow to restore coherence. Both are envelope violations. Both can occur independently. And both can interact.

Overshoot is the upward failure direction. It occurs when the displacement created by stress exceeds what the system can absorb without destabilizing. In prolonged fasting, overshoot may emerge through excessive cumulative load, progressive capacity mismatch, signaling amplification, or any combination of mechanisms that increase the effective size of the excursion beyond what the organism can contain. Once this happens, the system does not merely become "more challenged." It begins operating outside the recoverable stress range. This is where cascade initiation becomes possible. A local disruption can no longer remain local because the organism has already moved beyond the amplitude it can absorb cleanly.

Signaling distortion often intensifies here because once overshoot occurs, the communication layer of the system is no longer operating under conditions it can interpret proportionately. Disturbances begin propagating. Adjacent domains inherit altered conditions. Compensation becomes more active but less stabilizing. Instability spreads not simply because the original stressor was large, but because the excursion has carried the system above the point at which broad coherence can still be preserved. This is why overshoot is so important in the framework. It shows that instability can begin through excessive displacement even if recovery remains theoretically available. The excursion itself has become too large.

Undershoot is the downward failure direction. It occurs when recovery remains too shallow to restore coherence after displacement. This can happen even when stress amplitude itself is not dramatic in any single cycle. A system can remain below the upper bound repeatedly and still become unstable if it never returns deeply enough to reopen the room that stress has consumed. Recovery debt forms here because unresolved burden is carried forward. Margin compression develops because each incomplete return narrows the amount of ordinary variation the organism can tolerate. Cumulative constraint increases because the next cycle begins from a more constrained baseline than the last.

This is what makes undershoot so structurally important. It reveals that instability is not always loud at first. A system can appear relatively contained while still drifting outside the envelope through insufficient return. The problem is not that the body is failing to function in an immediate sense. The problem is that it is no longer restoring enough coherence between excursions to keep future excursions inside a recoverable range. Over time, this shallow return becomes just as destabilizing as overshoot because it progressively lowers the effective capacity of the system while preserving the appearance of continuity.

These two failure directions are distinct, but they also reinforce one another. Overshoot makes undershoot more likely because larger excursions are harder to recover from deeply. Undershoot makes overshoot more likely because shallower recovery leaves the system more constrained and therefore easier to displace beyond its upper bound on the next cycle. This interaction is one of the most important insights of the Oscillation Envelope Framework. Instability does not emerge from stress alone or from poor recovery alone. It emerges from envelope violation, and envelope violation can begin from either side. The organism may leave the recoverable range by going too high, by returning too little, or by doing both in sequence until the range itself becomes progressively harder to remain inside.

This is why the framework is not a recap of earlier mechanisms, but a unifying model. It does not merely restate prior chapters in simpler terms. It gives them a shared structure. Capacity, margin, recovery debt, failure cascade, cumulative load, and signaling distortion all matter because they either change the width of the envelope, increase the amplitude of oscillation, or reduce the depth of return. The envelope is what allows these mechanisms to be understood as components of one governing system rather than as separate physiological observations. Their common relevance is not thematic. It is geometric. They all determine whether the organism remains inside or outside the recoverable range.

Seen clearly, prolonged fasting is not simply a stress event. It is progressive pressure against the bounds of the oscillation envelope. As duration extends, the amplitude of displacement tends to increase while the conditions required for full return often become harder to maintain. The organism therefore moves closer to both forms of envelope violation at once. This is what makes prolonged fasting structurally consequential. It is not merely that the body is stressed for longer. It is that the body is progressively challenged to remain inside a bounded range where stress and recovery still remain recoverable together.

All of the prior mechanisms operate inside this framework because all of them ultimately affect either displacement, return, or the width of the recoverable range itself. Instability, in this model, is not defined by the existence of stress. It is defined by envelope violation. And prolonged fasting becomes important not because it introduces stress in isolation, but because it progressively increases pressure on both bounds of the system's recoverable operating range. System stability is determined not by avoiding stress, but by whether stress and recovery remain contained within a recoverable range.

Envelope Matching

Once the oscillation envelope has been established, a further question becomes unavoidable. If stability depends on remaining inside a recoverable range, what determines whether a given stressor will actually remain contained there? The answer is not stress alone, and it is not recovery alone. It is the relationship between them. A system does not remain stable simply because stress is present at a low or moderate level in abstraction, nor because recovery exists somewhere in the background as a general capability. Stability depends on whether the displacement created by stress is proportionate to the organism's current ability to absorb and resolve that displacement. This is the basis of the Envelope Matching Mechanism, or more specifically, the Amplitude Matching Model. It describes the requirement that stress magnitude must remain aligned with available recovery capacity if oscillation is to remain stable over time.

This matters because stress is often interpreted in isolation. It is evaluated as though its significance were contained entirely within its own magnitude, independent of the system receiving it. That interpretation is structurally incomplete. A stressor does not have a fixed meaning outside the organism in which it occurs. Its meaning depends on how much capacity the system has available, how deeply it can restore after displacement, and how much unresolved burden it is already carrying. The same amplitude can remain recoverable in one context and become destabilizing in another because stress is always relational. The organism does not respond to stress as an abstract quantity. It responds to whether that quantity can be fully integrated, resolved, and restored without narrowing coherence over time.

This is why the matching principle must be treated as a separate mechanism rather than as a mere restatement of the envelope. The envelope defines the recoverable range. Envelope matching defines the condition under which a given oscillation actually remains inside that range. It answers the operational question of stability. A system can have an envelope in theory, but if stress amplitude and recovery capacity are not proportionate, the organism will still begin drifting toward instability. Matching therefore functions as the dynamic requirement for stable oscillation. It is what allows the envelope to be preserved in practice rather than merely described in principle.

At the center of this mechanism is a simple but consequential clarification: stability requires alignment between displacement and return. Stress produces displacement. Recovery produces return. For oscillation to remain contained, the return must be adequate for the displacement that preceded it. This means stress amplitude must be matched not only by sufficient recovery depth but also by sufficient system capacity. Recovery depth matters because the organism must restore enough coherence after a given excursion for the next cycle not to begin from accumulated distortion. Capacity matters because the organism must have enough available bandwidth to absorb the excursion in the first place without destabilizing while it is still unfolding. Matching therefore does not refer to one variable alone. It refers to the proportional relationship between how far the system is moved and how much room and restoration it has available to contain that movement.

This is why stress is only stable when it can be fully absorbed and resolved. If the organism can absorb the excursion but not resolve it, instability still accumulates. If it can resolve some of the excursion but not absorb its full amplitude cleanly, instability also accumulates. Stability therefore requires a complete relation between the size of the perturbation and the system's total ability to process it. That relation is the heart of the Amplitude Matching Model. It moves the analysis away from absolute judgments such as "high stress" or "low stress" and toward the more structurally accurate question of whether a given stress is proportionate to what the system can currently recover from.

The importance of this relational framing becomes clearer when considering that the same stress does not produce the same outcome across systems. This is not because stress behaves inconsistently. It is because the systems receiving it differ in recovery depth, margin, capacity, and prior load. A moderate displacement in one organism may remain well matched to available restoration and therefore resolve cleanly. The same displacement in a narrower or more burdened system may exceed what can be fully absorbed or resolved, even if nothing about the stressor itself appears extreme. This is one of the central corrections the chapter must make. Moderate stress is not inherently stable. Stability depends on fit.

That fit can fail in more than one way. The first and most obvious is overmatch, where stress amplitude exceeds available recovery capacity. In this scenario, the excursion created by the stressor is too large relative to what the system can meaningfully restore. The organism is displaced farther than it can recover from without residual burden. This does not always require dramatic overload. It only requires enough amplitude relative to current capacity that the return becomes incomplete or destabilizing. Overmatch is therefore not defined by how severe the stress looks externally. It is defined by whether the system can actually resolve it fully enough to remain coherent afterward.

A second form of mismatch is undermatch, where recovery remains below the depth required for a given stress exposure. Here, the amplitude itself may not be extreme, but the return is too shallow. The organism is not restoring enough coherence to offset the displacement it has undergone. This means that even a seemingly manageable excursion can begin producing instability if recovery is insufficient relative to what that excursion demanded. Undermatch is particularly important because it reveals that mismatch is not always visible during the stress phase itself. The organism may appear to tolerate the event reasonably well while still failing to return deeply enough afterward to preserve stable oscillation over time.

A third form is dynamic mismatch, and this may be the most system-relevant of all. Dynamic mismatch occurs when stress and recovery begin aligned but drift out of proportion across repeated cycles or over the duration of a single prolonged event. In this case, neither stress nor recovery is necessarily extreme when examined in isolation. The instability emerges because the relationship between them changes. Stress amplitude may increase as cumulative load deepens, while recovery capacity simultaneously narrows through debt, compression, or signaling distortion. What was once matched becomes mismatched not because one variable suddenly becomes dramatic, but because the proportional relationship between them has shifted.

This dynamic quality matters because it explains why instability often emerges gradually rather than at the exact moment a stressor first appears significant. The system may begin in a condition where displacement and return are still relatively proportionate. But as duration extends or cycles accumulate, the amplitude of stress may begin consuming more of the system's available bandwidth while recovery becomes less complete. The relationship drifts. This is one of the reasons prolonged fasting becomes structurally consequential over time. It is not only that stress continues. It is that continued stress can progressively move out of alignment with the capacity required to resolve it.

The crucial clarification across all three mismatch types is that instability can emerge even when neither variable appears extreme alone. Stress can be moderate. Recovery can be present. Capacity can remain partially intact. And yet if the proportions between them are wrong, oscillation can still begin moving outside recoverable bounds. This is one of the central explanatory powers of the Envelope Matching Mechanism. It shows why instability is not reducible to dramatic overload or obvious deficiency. Misalignment itself is sufficient.

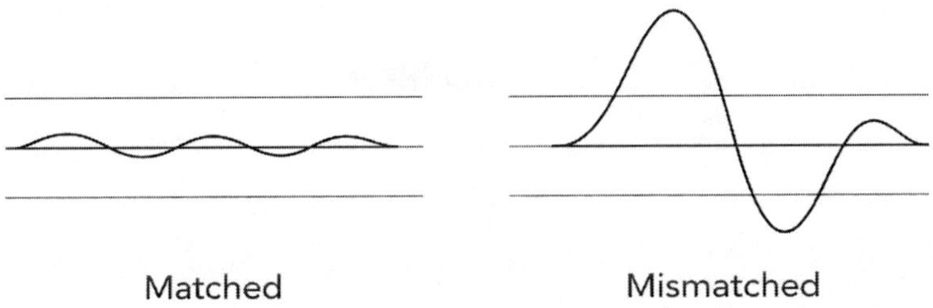

[Figure 16: Matched vs mismatched]

Once mismatch is present, instability outcomes begin to follow with increasing predictability. One of the first is overshoot. Overshoot occurs when the amplitude of displacement is too large relative to what the system can contain. In matched conditions, a stressor may produce movement without exceeding the system's ability to absorb it. In mismatched conditions, the same or only slightly larger excursion may now move above what the organism can stabilize. This is not simply a function of "too much stress" in the abstract. It is the consequence of stress no longer being proportionate to the recovery and capacity architecture available to support it.

Undershoot emerges from the opposite side of the same problem. When return is too shallow for the displacement that preceded it, the system fails to restore enough coherence to reset its recoverable operating space. It may remain below obvious acute instability, but it does not come back deeply enough to remain broadly matched over time. The result is that subsequent oscillations begin from a more constrained state. In this way, undershoot often appears quieter than overshoot initially, but it can be just as destabilizing structurally because it converts incomplete return into cumulative narrowing.

From here, several downstream consequences begin appearing as direct products of mismatch. Margin compression is one of the first. When stress and recovery are no longer aligned, the amount of buffer the system has available for ordinary deviation begins shrinking. This occurs because unresolved burden consumes the operating space that would otherwise protect the organism from small errors becoming consequential. The system becomes less forgiving not necessarily because the next stressor is larger, but because the room around it has narrowed. Mismatch therefore does not only create instability directly. It also reduces the margin that would have made future instability less likely.

Recovery debt follows naturally from the same process. If a given displacement is not fully matched by restoration, some portion of its burden remains unresolved. That unresolved portion is not neutral. It is carried forward into subsequent system states as residual imbalance. Recovery debt is therefore one of the accounting forms of mismatch. It represents the gap between what the stress required and what recovery delivered. Once this gap persists across cycles or deepens over time, the organism begins operating from a progressively more constrained baseline, making future matching harder to achieve even if the external stress remains unchanged.

Cascade risk also increases under mismatch because once displacement exceeds what can be absorbed or resolved, local instability becomes harder to contain. A mismatched system is more likely to allow disturbances to propagate because it lacks the proportional capacity and restoration depth required to localize them cleanly. The result is that what might once have remained a contained perturbation can begin altering adjacent systems through signaling, resource allocation, and regulatory dependency. This is why instability should not be understood as a function of magnitude alone. The real issue is whether the organism has enough matched structure to keep a disturbance from spreading.

This is the key principle that Chapter 18 must establish with precision: instability is the result of misalignment, not just magnitude. A stressor does not need to be maximal to become destabilizing. It only needs to be out of proportion to the system's available recovery depth and capacity. This is one of the reasons moderate stress can be structurally misleading. It may appear manageable when evaluated in isolation, but if it is applied to a narrowed, debt-carrying, poorly restoring system, its effective load becomes much larger than its abstract intensity would suggest. Misalignment changes meaning.

The final movement of the mechanism, then, is system alignment itself. Alignment should not be understood as a static condition or as a one-time achievement. It is a dynamic proportionality in which stress and recovery remain sufficiently matched across time for oscillation to stay contained. A system is aligned when displacement is proportionate to available capacity, return is deep enough relative to what was displaced, and the organism can continue cycling without progressively narrowing its own operating space. Alignment is therefore not the absence of challenge. It is the condition in which challenge remains structurally absorbable.

This dynamic quality matters because alignment must be maintained across cycles rather than assumed from one moment. A system may be aligned at the beginning of an event and drift out of alignment later as capacity narrows or cumulative burden rises. Conversely, a system may appear strained in a local moment while still remaining broadly aligned if the displacement is ultimately resolved in proportion to what it demanded. What matters is not any single point in isolation, but whether the relationship between amplitude and restoration remains coherent across time. This is what determines whether the organism stays inside the envelope.

The significance of alignment is therefore not merely that it supports stability in the present. It also preserves the conditions under which future stability remains possible. A matched system protects its own ability to continue oscillating recoverably because it does not spend more capacity than it can restore. A mismatched system progressively undermines its own future stability because each unresolved excursion narrows the room for the next one. This is why the Envelope Matching Mechanism belongs downstream of the envelope itself. The envelope defines the range; matching determines whether the system can keep living inside it.

Seen clearly, prolonged fasting becomes less a question of stress in abstraction and more a question of proportion. Stability does not depend on avoiding stress altogether. It depends on whether the amplitude of stress remains aligned with the depth of recovery and the capacity available to contain it. Mismatch can create instability even when stress appears moderate because the problem is not intensity alone. The problem is fit. Once displacement and return drift out of proportion, overshoot, undershoot, margin compression, recovery debt, and cascade risk begin emerging as predictable consequences of structural misalignment. System coherence is preserved only when stress amplitude and recovery capacity remain proportionate within the oscillation envelope.

Frequency vs. Magnitude Tradeoff

A stress event does not exist as a single point in time when viewed from the perspective of a living system. It becomes part of a pattern the moment it is repeated or extended across time. This is the starting condition for understanding the Frequency vs Magnitude Tradeoff and the reason the Tradeoff Curve Model is required as a distinct mechanism. The organism does not evaluate stress only by how large a single displacement is. It evaluates stress through how large that displacement is and how often similar displacements occur before the system has fully resolved them. Magnitude and frequency therefore cannot be separated without losing structural accuracy. They form a coupled relationship that determines whether oscillation remains contained or begins drifting toward instability.

Magnitude refers to the size of the displacement created by a stress event. It describes how far the system is moved away from its current state of coherence. In prolonged fasting, magnitude is shaped by the depth of depletion, the degree of signaling alteration, the extent of downregulation, and the accumulation of constraint that develops as duration extends. Frequency refers to how often that displacement is imposed. It describes the temporal spacing between stress events, whether those events are discrete or continuous, and whether sufficient time exists between them for the system to restore coherence. These two variables are not independent. They are linked through the system's capacity to absorb and resolve stress within a bounded range.

The core principle of the Tradeoff Curve Model follows directly from this relationship: higher magnitude requires lower frequency to remain stable. This is not a recommendation or a preference. It is a structural requirement imposed by the finite nature of capacity, recovery depth, and margin. A system can sustain repeated stress if each event is small enough relative to its ability to absorb and resolve it. A system can also sustain large stress events if they occur infrequently enough for recovery to fully restore coherence before the next event. What it cannot sustain is the pairing of large displacement with rapid repetition. High magnitude and high frequency together exceed the organism's ability to remain inside a recoverable range because they compress both sides of the oscillation simultaneously: they increase the size of each excursion and reduce the time available to return from it.

This is why stress events cannot be evaluated in isolation. A single prolonged fast may appear structurally containable when considered alone, particularly if the system begins from a relatively broad capacity state. But once that event is placed into a pattern of repetition, its meaning changes. The organism is no longer managing one displacement. It is managing a sequence of displacements whose cumulative effect depends on how fully each one is resolved before the next begins. Repetition therefore interacts with amplitude. It determines whether the system is allowed to re-expand after stress or whether it is forced to operate under progressively constrained conditions as unresolved burden accumulates.

The tradeoff is continuous rather than binary. There is no single threshold at which frequency becomes acceptable or unacceptable independent of magnitude. Instead, the relationship forms a curve. At one end of the curve, low-magnitude stress can be repeated more frequently because each displacement remains small relative to available capacity and can be resolved more quickly and completely. At the other end, high-magnitude stress requires longer intervals between events because each displacement consumes more capacity, requires deeper restoration, and produces greater potential for residual distortion if recovery is incomplete. Between these ends lies a continuum of combinations, each with its own implication for whether oscillation remains contained.

This curve is not abstract. It is shaped by the system's current condition. Recovery capacity influences how quickly and how completely the organism can restore coherence after displacement. A system with broad recovery capacity can move further along the frequency axis for a given magnitude because it can resolve each event more effectively before the next one occurs. Margin influences how much deviation can be tolerated without producing disproportionate consequences. A system with broader margin can absorb repeated perturbations more quietly because small errors remain small relative to available buffering space. Cumulative load influences how much of the system's capacity is already occupied by unresolved strain. A system carrying high load begins each event from a more constrained baseline, effectively shifting its position on the curve toward lower allowable frequency even if the external magnitude remains unchanged.

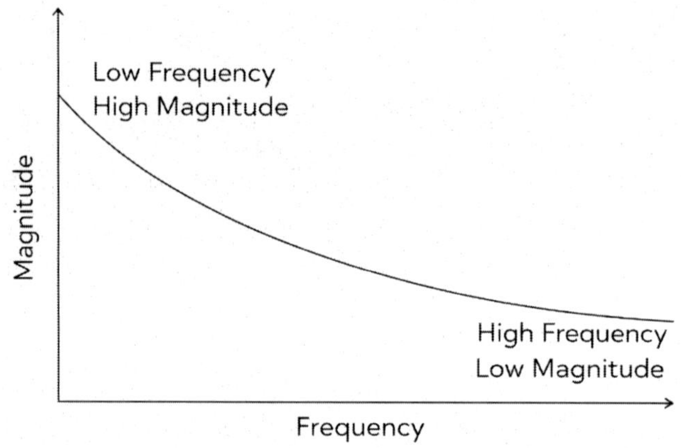

[Figure 17: Tradeoff curve-frequency]

The curve can therefore be understood as a boundary separating stable and unstable combinations of magnitude and frequency. Combinations that fall under the curve remain inside the system's ability to absorb and resolve stress over time. Combinations that rise above the curve exceed that ability. The distinction is not about whether stress exists. It is about whether the pattern of stress remains proportionate to the system's recovery architecture. This is why the curve must be treated as structural rather than optional. The organism cannot choose to ignore it. If magnitude and frequency are paired in a way that exceeds the system's ability to resolve them, instability emerges regardless of how the stress is framed conceptually.

Misuse of this relationship often occurs through pattern-level errors rather than through obviously extreme single events. One of the most common is the stacking of high-magnitude events too closely together. Each event in isolation may appear within the system's capacity when considered without reference to time. But when repeated before recovery has fully restored coherence, the system is forced to manage multiple unresolved displacements simultaneously. The effective magnitude of the pattern becomes larger than the magnitude of any individual event because the organism is no longer starting from a restored baseline. This is how repetition amplifies load. It converts what might have been contained stress into cumulative burden.

Another form of misuse emerges from the assumption that recovery scales automatically with stress. This assumption treats recovery as a passive or guaranteed response, as though the system will simply expand its restoration capacity to match whatever magnitude or frequency of stress is imposed. In reality, recovery is constrained. It depends on signaling coherence, resource availability, autonomic regulation, and the absence of competing demands. When magnitude increases or frequency rises, recovery does not necessarily expand in parallel. Instead, it may become less complete or slower, particularly if cumulative load and margin compression are already present. The result is a progressive drift above the tradeoff curve even when no single event appears extreme.

A third pattern error involves treating stress events as independent. This occurs when each exposure is evaluated as though it begins from a neutral baseline unaffected by prior events. In a living system, this assumption does not hold. Each event modifies the state from which the next one begins. If recovery is incomplete, the system carries forward residual burden. If margin has been compressed, subsequent perturbations occupy a larger proportion of the available range. If signaling has been distorted, coordination becomes less precise. Treating events as independent obscures the way repetition transforms system load. It hides the cumulative nature of the tradeoff and makes instability appear sudden when it is actually the product of pattern misalignment over time.

These misuse patterns converge on a single structural insight: instability often emerges from the interaction between magnitude and frequency rather than from the absolute size of any one stressor. A moderate event repeated too frequently can become destabilizing even if each instance appears manageable. A large event spaced appropriately can remain contained if the system is given sufficient time and capacity to restore coherence. The system is not responding to intensity alone. It is responding to the pattern in which that intensity is embedded. This is why the Tradeoff Curve Model belongs at the pattern level of analysis. It explains behavior that cannot be understood by examining events in isolation.

The implications for stability follow directly. To remain within a recoverable range, the organism must operate under the curve defined by its current capacity, recovery depth, and margin. This does not require the elimination of stress or the minimization of magnitude to trivial levels. It requires that the pairing of magnitude and frequency remains proportionate to what the system can resolve. When this condition is met, repeated stress can occur without progressive narrowing because each cycle allows sufficient restoration. The system is displaced and then returned in a way that preserves its ability to do so again.

Frequency plays a decisive role here because it determines whether recovery is allowed to complete. A system can tolerate a given magnitude only if enough time exists for restoration to reach the depth required to reset the recoverable range. When frequency increases beyond what recovery can support, even moderate stress begins to accumulate as unresolved burden. Recovery debt forms not because the stress is inherently excessive, but because the system is not given enough time to resolve it fully. This is why time structure governs stability as much as magnitude. It controls whether restoration is complete or partial, whether margin is preserved or compressed, and whether capacity is maintained or progressively consumed.

This also explains why increasing frequency can transform the meaning of a stressor without changing its external characteristics. A displacement that remains contained when it occurs occasionally can become destabilizing when repeated rapidly because the system's starting point shifts. Each event begins from a less restored state than the last. The effective amplitude increases because the organism is already partially displaced when the next event arrives. This is how repetition alters system load. It changes not only how often stress occurs, but how much of that stress remains unresolved at any given time.

The relationship between magnitude and frequency is therefore inseparable from the system's position relative to the oscillation envelope. Operating under the tradeoff curve allows oscillation to remain contained because each displacement is matched by sufficient recovery within the available time. Operating above the curve pushes the system toward envelope violation because either the amplitude is too large, the frequency is too high, or both are misaligned with the organism's capacity to resolve them. Instability in this context is not the result of a single failure. It is the outcome of a pattern that the system cannot sustain.

The Tradeoff Curve Model also clarifies why instability can appear delayed. A system may initially tolerate a given combination of magnitude and frequency because capacity and margin are still relatively broad. But as cumulative load increases and recovery becomes less complete, the same pattern can move above the curve without any obvious change in the external stress. The organism has shifted. The curve has effectively moved downward due to narrowing capacity, reduced margin, and growing recovery debt. What was once a stable pairing becomes unstable because the system's ability to resolve it has changed. This is why ignoring cumulative effects of repetition leads to misinterpretation. It treats stability as fixed when it is actually state-dependent.

Seen clearly, magnitude and frequency form a coupled constraint that governs whether repeated stress remains recoverable. Larger events require longer recovery and therefore lower allowable repetition frequency. Smaller events can be repeated more frequently because they consume less capacity and can be resolved more quickly. Between these extremes lies a continuous tradeoff shaped by the system's current condition. Instability emerges when this relationship is ignored, when high magnitude is paired with high frequency, or when repetition is increased without regard for the system's ability to restore itself between events.

System stability therefore depends not only on how large stress is, but on how often it is repeated relative to the system's ability to recover within the oscillation envelope.

Strategic Use

A system does not treat all inputs as equivalent simply because they share a common label. The same input can function as a stabilizing influence in one context, a neutral perturbation in another, and a destabilizing force in a third, depending on how it interacts with the organism's current state. Prolonged fasting is one such input. It is often framed as though it represents a generalizable condition that can be applied across contexts without structural distinction. That framing obscures the mechanism that actually governs its effects. Prolonged fasting is not a baseline operating condition of the system. It is a discrete stressor whose viability depends on the state into which it is introduced. This is the basis of the Strategic Use Mechanism, or more precisely, the Conditional Deployment Model. It establishes that prolonged fasting must be understood as a situational input whose effects are determined by capacity, recovery alignment, and cumulative system condition, rather than as a continuously applicable or inherently stabilizing state.

This distinction begins with the recognition that the organism does not exist in a single static condition across time. It moves through changing levels of capacity, margin, recovery depth, and accumulated burden. Any input introduced into the system therefore interacts with a dynamic structure rather than a neutral baseline. Prolonged fasting, when viewed through this lens, cannot be treated as a default pattern. It imposes a specific type of displacement characterized by increasing amplitude, evolving substrate conditions, signaling shifts, and cumulative constraint. That displacement is meaningful only in relation to the system's ability to absorb and resolve it. If the organism has sufficient capacity, adequate margin, and the ability to restore coherence after the event, the displacement may remain contained within the oscillation envelope. If those conditions are not present, the same input can produce progressive instability even if its external characteristics remain unchanged.

This is why conditional use must be defined at the outset. Prolonged fasting is viable only when it remains compatible with the system's current envelope. Compatibility here does not refer to preference or tolerance in a subjective sense. It refers to whether the amplitude of the event, its interaction with cumulative load, and its impact on recovery remain proportionate to the organism's available capacity. A system with broader capacity, lower accumulated burden, and more complete recovery processes may absorb the same fasting event that a more constrained system cannot. The input itself does not change. The system receiving it does. This is the central logic of the Conditional Deployment Model: viability is determined by context, not by the inherent properties of the input alone.

Once this principle is established, the boundary conditions that govern deployment become clearer. Capacity is one of the primary constraints because it defines how much displacement the system can absorb without losing coherence. A prolonged fast increases the magnitude of displacement over time through depletion and cumulative strain. If that magnitude approaches or exceeds what the system can contain, the event no longer remains inside the envelope. The issue is not whether the fast is conceptually valid. It is whether the system has enough available bandwidth to support the displacement it creates. Capacity is therefore not an abstract limit. It is a dynamic boundary that determines whether the event remains structurally viable at the time it occurs.

Recovery requirements form a second boundary condition. The system does not maintain stability simply by absorbing stress. It must also restore what that stress consumes. Prolonged fasting interacts directly with recovery because it alters substrate availability, signaling conditions, and output regulation in ways that must be re-synchronized afterward. If recovery depth is insufficient relative to the displacement created by the fast, unresolved burden accumulates. This accumulation narrows the system's operating range and reduces its ability to remain stable across subsequent cycles. Viability therefore depends not only on whether the fast can be completed, but on whether the organism can restore coherence after it in a way that preserves its capacity for future oscillation.

Margin availability represents a third constraint. Margin is the buffer that allows the system to tolerate imperfect transitions and small mismatches without destabilizing. Prolonged fasting consumes margin by increasing sensitivity, altering signaling precision, and compressing the room for deviation. A system with broad margin can absorb this compression more quietly because it has excess buffering space. A system with narrow margin cannot. In that context, even moderate displacement can become structurally significant because it occupies a larger proportion of the available tolerance band. Margin therefore determines how forgiving the system is to the stress introduced by fasting and whether ordinary variation during the event can remain contained.

Cumulative load integrates these factors across time. It reflects the total burden the system is already carrying through unresolved stress, recovery debt, signaling distortion, and prior constraint accumulation. A system under low cumulative load begins a prolonged fast with more available capacity, more margin, and a broader ability to recover. A system under high cumulative load begins from a constrained baseline. The same fasting event therefore has a different structural meaning depending on the level of load already present. Cumulative load does not change the external characteristics of the fast, but it changes how much of the system's remaining bandwidth that fast consumes.

Together, these boundary conditions define whether prolonged fasting remains inside or outside the oscillation envelope. They determine whether the displacement created by the event can be absorbed and resolved or whether it will exceed capacity, compress margin, and generate recovery debt. This is why deployment must be understood as bounded rather than open-ended. The system cannot sustain indefinite or unconstrained application of a stressor simply because that stressor is conceptually valid. Once the input begins interacting with capacity limits, recovery constraints, and cumulative burden in a way that exceeds the envelope, instability follows as a structural outcome.

This boundedness becomes more evident when prolonged fasting is placed within the broader system in which it occurs. The fast does not exist independently. It interacts with recovery cycles, external load, and the organism's evolving state over time. Sleep quality, environmental stress, activity demands, and prior perturbations all shape how the system receives and processes the fasting event. The organism does not isolate the fast from these factors. It integrates them into a single regulatory field. This means that the viability of the fast cannot be evaluated in isolation. It must be evaluated in terms of how it fits into the total system behavior at the time it occurs.

This integration also clarifies that prolonged fasting is one input among many. It does not define the system by itself. It participates in a network of interactions that include recovery processes, external demands, and internal regulatory conditions. When the fast is introduced into this network, it modifies the conditions under which other processes must operate. It alters resource allocation, shifts signaling priorities, and changes the timing and depth of recovery. These changes do not remain confined to the fasting event. They influence the entire system context in which the organism continues to function. This is why the Conditional Deployment

Model emphasizes fit rather than isolated effect. Stability depends on how the input integrates with the whole, not on the input alone.

From this perspective, the non-escalation principle becomes structurally necessary. Increasing the duration or intensity of prolonged fasting does not inherently improve outcomes because the system is bounded. Escalation increases the amplitude of displacement. It moves the system further away from its current state and therefore increases the demands placed on capacity and recovery. At the same time, larger amplitude reduces the allowable frequency at which the event can be repeated without generating instability. This is consistent with the tradeoff relationship between magnitude and frequency. As amplitude rises, the system requires more time and more restoration to return to coherence. Escalation therefore narrows the conditions under which the fast can remain inside the envelope.

Escalation also compresses margin. As displacement increases, the system has less buffer for absorbing mismatch and variation. This makes the organism more sensitive to deviations during and after the event. Small errors become more consequential because they occupy a larger proportion of the remaining tolerance band. The result is that higher intensity does not simply increase the magnitude of the stress. It also reduces the system's ability to tolerate imperfection while that stress is unfolding. This dual effect makes escalation structurally significant. It increases the demands placed on the system while simultaneously reducing the system's ability to manage those demands.

These effects demonstrate why more stress does not equal more benefit in a bounded system. The organism does not respond to increased intensity with unlimited adaptive gain. It responds within the constraints of capacity, recovery, and margin. Once those constraints are approached or exceeded, additional stress produces diminishing returns and increasing instability. This is not a matter of preference or interpretation. It is a consequence of operating within a finite system. The boundaries imposed by the oscillation envelope cannot be bypassed through escalation. They can only be respected or violated.

The failure modes associated with conditional deployment arise when these constraints are ignored. One of the most common is treating prolonged fasting as universally applicable. This assumes that the input has a consistent effect regardless of system state. In reality, the same fasting event interacts differently with different levels of capacity, recovery, and cumulative load. Applying it without regard to these variables leads to mismatches between displacement and the system's ability to resolve it. The result is not uniform adaptation, but variable outcomes shaped by the underlying state of the organism.

Another failure mode involves escalating duration without regard to capacity. This treats the system as though it can absorb increasing amplitude indefinitely. As duration extends, however, the demands placed on capacity, recovery, and margin also increase. If those demands exceed what the system can support, instability emerges. The issue is not that duration itself is inherently problematic. It is that duration alters the magnitude of displacement in a way that must remain proportionate to the system's constraints. Ignoring this relationship leads to structural mismatch.

A third failure mode arises from ignoring system state and cumulative load. This assumes that each fasting event begins from a neutral baseline unaffected by prior stress. In practice, the system carries forward unresolved burden, altered signaling, and reduced margin from previous perturbations. Introducing a new stressor into this context changes its effective load. What appears moderate in isolation may become significant when layered onto existing constraint. This is why cumulative load must be considered as part of deployment. It determines how much of the system's remaining bandwidth the new event will consume.

These failure modes share a common feature. They treat prolonged fasting as a default input rather than a conditional one. They assume that the input can be applied without regard to context, timing, or system condition. The Conditional Deployment Model corrects this by placing the input within the constraints that actually govern its effects. It reframes fasting as a bounded stressor whose viability depends on its compatibility with the system's current state. This does not assign value to the input in isolation. It assigns structure to how the input interacts with the organism.

When viewed through this model, prolonged fasting becomes neither inherently stabilizing nor inherently destabilizing. It becomes conditionally viable. Its effects are determined by whether it remains inside the limits imposed by capacity, recovery, margin, and cumulative load. If those limits are respected, the displacement it creates can be absorbed and resolved. If they are exceeded, the same displacement contributes to mismatch, recovery debt, margin compression, and potential cascade. The difference lies not in the input itself, but in the conditions under which it is deployed.

This framing also reinforces that stability is not achieved by eliminating stress. The system requires movement between stress and recovery to remain functional. What matters is whether that movement remains contained. Prolonged fasting is one form of movement within this broader system. Its role is defined by how it fits into the total pattern of displacement and return. When it aligns with the system's capacity and recovery architecture, it remains part of a stable oscillation. When it does not, it becomes a source of instability not because it is intrinsically harmful, but because it has been introduced outside the bounds that allow it to be resolved.

Prolonged fasting therefore cannot be understood as a default state of operation. It is a bounded, conditional stress input whose viability depends on its alignment with the system's constraints. Its effects are determined by how it interacts with capacity, recovery, margin, and cumulative load at the time it is introduced. System stability depends on whether fasting is deployed within the constraints of capacity, recovery, and the oscillation envelope, rather than assumed to be universally or continuously applicable.

Strategic Non-Use

A bounded system does not become unstable only because a stressor is large. It also becomes unstable when that stressor is introduced into a state that no longer has the structural room to contain it. This distinction is essential because it separates the question of whether prolonged fasting can function as a viable input from the question of whether it is conceptually meaningful in isolation. A stressor may remain valid as a mechanism and still become non-viable as an application if the system receiving it no longer has enough capacity, recovery depth, or coherence to absorb its displacement. This is the basis of the Non-Use Mechanism, or more specifically, the Non-Deployment Condition Model. It defines the conditions under which prolonged fasting becomes structurally incompatible with stability. In this model, non-deployment is not a matter of preference, philosophy, or behavioral reluctance. It is the direct consequence of system state.

The importance of this distinction becomes clear once it is recognized that not all non-use is elective. Some states simply do not support additional stress without violating the conditions required for contained oscillation. A system can only sustain displacement if it has sufficient bandwidth to absorb that displacement and sufficient recovery depth to resolve it. When those conditions are no longer present, the viable range for introducing further stress narrows or disappears entirely. The issue is not whether the stressor is generally useful or generally harmful. The issue is whether the organism can currently receive it without moving further outside the oscillation envelope. Once that possibility is no longer structurally available, non-deployment emerges as the only coherent state description.

Capacity is the first of the governing constraints. In systems terms, capacity is the finite, state-dependent bandwidth through which stress can be absorbed while preserving coherence. It is not a fixed trait, nor a vague reference to "tolerance." It is the total functional room the organism has available at a given moment to manage displacement without destabilizing. This room includes metabolic flexibility, signaling coherence, recovery reserve, and margin. When capacity is broad, the system can absorb perturbation without immediate structural distortion. When capacity is heavily consumed, the organism may still be functioning, but it is doing so with less available room to accommodate further load.

This consumed state matters because prolonged fasting does not enter the system neutrally. It creates a displacement. If capacity is already occupied by unresolved burden, compensatory effort, or narrowed margin, there may no longer be enough uncommitted bandwidth for that displacement to remain contained. Additional stress does not simply layer on top of a stable baseline. It enters a system already operating near or within its tolerance boundaries. The same fasting event that might have remained viable in a broader-capacity context can therefore become incompatible once capacity has been sufficiently consumed. This is why insufficient capacity does not merely reduce flexibility. It can eliminate viable deployment range altogether.

Limited margin is one of the clearest expressions of this condition. Margin is the local buffering space that allows the system to tolerate deviation without translating that deviation into instability. When capacity is heavily consumed, margin often narrows with it. The organism may still preserve outward continuity, but it does so with less room for error, less tolerance for mismatch, and less ability to absorb additional displacement without consequence. In such a state, the system becomes more sensitive not because the next stressor is inherently extreme, but because the room around it has diminished. Additional stress now occupies a larger proportion of what remains.

Reduced flexibility reinforces the same constraint from another angle. A system with broad flexibility can redistribute load, shift across substrate conditions, and preserve coherence under changing demands. A system with reduced flexibility must regulate more tightly. It has fewer available pathways through which to adapt to perturbation. This means that prolonged fasting no longer interacts with a broad adaptive field. It interacts with a narrowed one. The displacement it creates becomes harder to distribute, harder to localize, and harder to resolve. The consequence is not simply greater difficulty. It is reduced structural compatibility.

Increased sensitivity follows naturally from both of these changes. Once capacity is consumed and flexibility narrows, the organism becomes more reactive to additional load because less of that load can be quietly buffered. Small mismatches become more consequential. Ordinary perturbations begin producing larger effects. This is one of the reasons additional stress becomes structurally incompatible in high-consumption states. The system is not simply "less robust" in a generalized sense. It is operating with less available room to absorb the very type of displacement that prolonged fasting imposes. Under these conditions, the capacity required for viable deployment is no longer present.

Recovery limitations create a second major condition of incompatibility. Recovery is the mechanism through which the system restores coherence after stress. It reopens operating space, re-synchronizes signaling, replenishes reserve, and reduces residual distortion. Without sufficient recovery, the organism cannot maintain stable oscillation over time because each displacement leaves behind unresolved burden that narrows the starting point for the next cycle. Recovery is therefore not secondary to viability. It is one of the primary conditions that determines whether any additional stress can remain compatible with stability.

When recovery is delayed, the system remains in a partially displaced state for longer. The organism does not fully return to a more coherent baseline before the next load arrives or before the residual effects of the prior load have been resolved. This delay matters because it means that even if the system has not acutely destabilized, it is still carrying the structural consequences of prior displacement forward. Prolonged fasting introduced into such a state does not enter a restored organism. It enters one that is still partially compressed. The new displacement therefore interacts not with a broad recoverable range, but with a system already narrowed by incomplete return.

Incomplete recovery intensifies this condition further. A system may appear to move back toward baseline after stress while still failing to restore the depth of coherence required for stable oscillation. Signaling may remain narrowed, margin may remain compressed, or adaptive flexibility may remain reduced even after the overt stress phase has ended. In this state, the organism has not fully resolved the prior displacement. It has merely stabilized at a lower level of available room. Additional stress introduced here is therefore not additive in a neutral sense. It compounds unresolved strain because the system is still carrying the structural debt of prior perturbation.

Compressed recovery represents the same problem in temporal form. The organism may have some capacity to restore itself, but not enough time, depth, or available room to do so fully before further load is imposed. Recovery is therefore not only incomplete; it is structurally constrained. The system cannot re-expand after displacement because the conditions required for that expansion are absent or insufficient. This is why recovery limitation is such a decisive non-deployment condition. It prevents the organism from resolving what it has already absorbed. Once that mechanism is impaired or compressed, stable oscillation becomes increasingly impossible.

The consequences of this limitation are cumulative. Accumulation follows because unresolved burden remains in the system instead of being cleared. Residual signaling distortion, narrowed margin, and incomplete substrate reorganization all become part of the baseline from which future stress must be managed. Unresolved strain is therefore not an abstract concept. It is the physical and regulatory burden left behind when recovery fails to restore coherence. Under these conditions, prolonged fasting no longer functions as a discrete perturbation that can be integrated and resolved. It becomes a new load imposed onto an organism already unable to fully process the last one. This is why insufficient recovery prevents stable oscillation. The system cannot return deeply enough to make further displacement structurally compatible.

System instability forms the third major condition. Instability here should not be understood as overt collapse or obvious dysfunction alone. It refers more broadly to a state in which the organism is already outside or pressing closely against the oscillation envelope. The system may be reactive, misaligned, or carrying enough cumulative load that ordinary perturbations are no longer being contained cleanly. In such a state, prolonged fasting does not enter a stable field. It enters one in which the architecture of containment is already compromised.

Misalignment is one expression of this condition. A system can remain active while still being out of proportion internally. Signaling, output, and recovery may no longer be synchronized in a way that preserves broad coherence. The organism may still be functioning, but it is doing so through compensation rather than through stable containment. Additional stress introduced here does not interact with a well-organized baseline. It interacts with an already unstable coordination structure. This makes further displacement more likely to propagate rather than remain local.

Reactivity is another expression. Once the system is operating with narrowed margin and heightened sensitivity, even moderate perturbations begin producing larger downstream consequences. This does not require catastrophic instability to already be present. It only requires enough compression that the organism can no longer buffer new load quietly. Prolonged fasting introduced into such a state is therefore not merely "one more stressor." It is a perturbation entering a system already primed to amplify disturbance rather than absorb it. This is why unstable systems cannot absorb further displacement. Their operating conditions have already shifted from containment toward propagation.

Cumulative load completes the picture. A system carrying substantial unresolved burden is structurally different from one beginning from a broad, low-load state. The same prolonged fast therefore has a different meaning depending on how much load the organism is already carrying. In a high-load state, additional stress increases the probability that local instability will spread because there is less capacity, less margin, and less recovery reserve available to localize it. Cascade risk rises not because the new stressor is necessarily extreme, but because the system into which it is introduced is already operating too close to its limits to contain it cleanly.

This is where non-deployment must be understood as a structural outcome rather than a behavioral principle. It is not a decision rule in the abstract, and it is not guidance about what ought to be done. It is simply the name for the state in which prolonged fasting no longer fits within the system's current constraints. The organism has moved into a condition where the displacement created by the fast would exceed available capacity, violate recovery requirements, or intensify existing instability. Under those conditions, the input becomes structurally incompatible with stability. Non-use is therefore not elective in any meaningful systems sense. It is what remains when viable deployment range has been consumed.

This distinction matters because it prevents the mechanism from being confused with preference or caution. The issue is not whether prolonged fasting is "desirable" under these conditions. The issue is that the system no longer has the architecture required to support it. Once capacity is insufficient, recovery is limited, and instability is already present, the consequences of introducing further stress become predictable. Overshoot becomes more likely because the additional displacement exceeds what the organism can absorb. Undershoot becomes more likely because the system cannot restore coherence deeply enough after that displacement. Cascade becomes more likely because local disturbances are no longer being contained within a stable field.

These outcomes are not separate failure modes so much as different expressions of the same incompatibility. A system with insufficient capacity cannot contain the amplitude. A system with insufficient recovery cannot resolve the displacement. A system already unstable cannot localize the consequences. The result is that prolonged fasting ceases to function as a conditionally viable stressor and becomes structurally non-viable instead. This is not because the mechanism of fasting has changed. It is because the state of the organism has changed in a way that removes the conditions required for contained oscillation.

Seen clearly, the Non-Deployment Condition Model does not add a new moral layer to the discussion. It simply closes the structural logic of the system. Prolonged fasting is not universally applicable because not all system states can absorb additional stress while preserving coherence. Capacity can be too consumed, recovery can be too limited, and instability can already be too active for further displacement to remain inside the envelope. When those conditions are present, non-use is not a preference or a strategy. It is the direct consequence of structural incompatibility. Prolonged fasting becomes structurally non-viable when system capacity, recovery, and stability conditions no longer support contained oscillation within the envelope.

Magnitude Without Structure

Duration is often granted a meaning it does not possess. Once a stressor extends across time, it is frequently interpreted as though extension itself confers significance. Longer is taken to imply deeper, stronger, more meaningful, or more effective. In this framing, magnitude begins to function as a proxy for value. The larger the stress, the more consequential it is assumed to be. But this assumption does not hold at the level of system behavior. Magnitude, whether expressed as duration, intensity, or depth, is structurally neutral. It describes how much stress is being applied. It does not determine what that stress will produce. A larger perturbation is not inherently more useful, more stabilizing, or more effective than a smaller one. It is only larger. That distinction is the central correction required to understand prolonged fasting clearly. Magnitude without structure produces instability.

This matters because stress does not carry its own meaning independently. It only acquires meaning when placed inside the conditions that determine whether it can be absorbed, resolved, and contained. Without those conditions, duration is merely extension. Intensity is merely displacement. Depth is merely amplitude. None of these variables predicts outcome on its own. A longer fast does not inherently produce a better or more coherent result than a shorter one. A deeper stress event does not inherently generate more useful adaptation than a contained one. Magnitude alone has no governing power because systems do not respond to size in abstraction. They respond to whether the size of a perturbation is proportionate to what the organism can structurally support.

The determinant is therefore not stress itself, but structure. Structure is the relationship between stress amplitude, recovery depth, system capacity, margin, timing, and accumulated load. It is the architecture that gives stress a consequence. When structure is intact, a stressor can be absorbed and resolved without destabilizing the system that receives it. When structure is absent, distorted, or exceeded, the same stressor produces a different outcome because the organism no longer has the conditions required to contain it. Structure is what determines whether a perturbation remains local or propagates, whether recovery restores coherence or leaves debt, whether the system returns broadly or narrows further after displacement. Outcome is not a direct property of stress. It is an expression of the structure into which stress is introduced.

This is why duration cannot be treated as a governing variable by itself. Duration matters only because it alters the relationship between amplitude and the structure required to support it. As a stress event extends, it deepens depletion, increases constraint, narrows margin, and places greater demands on recovery. None of these effects are governed by time alone. They are governed by how time interacts with the system's ability to contain what is unfolding. The same duration can remain recoverable in one state and become destabilizing in another because duration has no standalone authority. It only matters through the structural conditions that interpret it.

The same is true of intensity more broadly. Systems do not reward excess simply because it is excess. They do not recognize extension as an achievement or escalation as a sign of deeper value. They register whether the perturbation remains compatible with the architecture of containment. If the stress amplitude exceeds what recovery can resolve, instability begins. If capacity is already consumed, the same magnitude becomes more consequential. If margin has narrowed, small deviations become larger in effect. If cumulative load is already present, additional displacement no longer behaves as an isolated event. Structure determines whether stress remains absorbable or becomes destabilizing. The magnitude of the stressor alone cannot answer that question.

This is why the full system resolves into a single organizing condition: whether the organism remains inside the oscillation envelope. The envelope does not need to be restated in full because its function is already clear. It is the bounded range within which stress and recovery can occur without loss of coherence. Every mechanism developed throughout the system ultimately converges here. Cumulative load matters because it narrows the room available inside the envelope. Recovery debt matters because it prevents return from fully restoring that room. Margin compression matters because it reduces the tolerance for deviation while still inside the range. Capacity constraints matter because they determine how much displacement the organism can absorb before crossing its upper boundary. Rebound instability matters because even the return phase can become destabilizing if it exceeds or fails to meet what the system can structurally resolve. All of these mechanisms differ in form, but not in final logic. They describe ways in which structure can fail to contain stress and recovery within a recoverable range.

Seen from this perspective, instability is not situational in the superficial sense. It is not a random consequence of isolated events or an accidental byproduct of intensity alone. It is structural. It emerges when magnitude exceeds what structure can support. It appears when recovery no longer matches stress, when capacity is exceeded, when margin is lost, or when cumulative burden consumes the system's available room for containment. The surface expression may vary, but the governing principle does not. Instability is the result of stress no longer being held inside the architecture required to absorb it.

This is also where the central error becomes fully visible. The primary mistake is not merely applying too much stress. It is treating magnitude as though it were a controllable source of benefit while ignoring the structural limits that determine whether that stress can be integrated. This error appears whenever duration is interpreted as proof of effect, whenever escalation is mistaken for depth, or whenever extension is treated as inherently meaningful independent of the system receiving it. These interpretations assume that systems respond positively to "more" as a general rule. They do not. Systems respond to alignment, containment, and recoverability. Excess does not create coherence. Escalation does not create stability. Extension does not create benefit. If structure is absent, each of these can accelerate instability rather than deepen adaptation.

That is because biological systems are not organized around the reward of maximal stress. They are organized around the preservation of coherent function under variable load. What they can use is not defined by what can be imposed, but by what can be absorbed and resolved without loss of organization. This is the difference between a stressor and a structured input. A stressor becomes structurally useful only when it remains proportionate to the system's ability to contain it. Once that proportion is lost, the same input begins producing debt, compression, mismatch, or cascade rather than stable adaptation. The error is therefore not simply excess in quantity. It is the assumption that quantity can substitute for structure.

Everything in the system can be compressed into a single logic chain: stress, structure, outcome. Stress is the perturbation. Outcome is the result. But structure is what determines the relationship between them. Without structure, stress has no reliable meaning. It can destabilize, propagate, compress, or accumulate. With structure, the same stress can remain bounded, recoverable, and coherent. This is the real governing law. Stress does not determine outcome directly. Structure determines whether stress becomes adaptive or destabilizing.

That compression matters because it removes the illusion that the system is governed by effort, extremity, or extension. It is not. It is governed by whether displacement remains proportionate to the architecture required to resolve it. Duration does not create benefit. Magnitude does not create stability. Only structure determines whether stress can be absorbed without destabilization. Prolonged fasting is not governed by how much stress is applied, but by whether that stress remains structured within the system's capacity to recover, contain, and remain coherent over time.

About the Author

Nathan Reed writes about metabolic stability in systems exposed to load, variability, and time.

His work examines how biological capacity is shaped by energy availability, recovery architecture, sleep amplitude, muscle mass, environmental demand and workload. Rather than treating nutrition, fasting, sleep, and training as isolated tactics, he approaches them as interacting regulatory systems governed by margin, buffering, oscillatory coherence, and constraint dominance.

The Stability framework explores how physiological compression accumulates gradually and how proportional calibration preserves long-horizon resilience across decades. The objective is not peak optimization, but preservation of functional continuity under load.

Prolonged Fasting is part of a broader metabolic stability ecosystem developed under Perennial Publishing, each volume addressing a distinct vector of biological durability.

The work is designed for readers who prefer structural clarity over motivational rhetoric and long-horizon coherence over short-term intensity.

The Metabolic Stability Series

Metabolic Stability: *A Systems Guide to Blood Sugar, Fasting, and Long-Term Metabolic Health*

(Cornerstone Volume)

Fasting Fundamentals: *A Systems Guide to Metabolic Flexibility, Fat Loss, and Long-Term Health*

(Foundational Volume)

Intermittent Fasting: *Metabolic Timing, Body Composition, and Sustainable Health*

Prolonged Fasting: *A Systems Guide to Duration, Recovery, and Metabolic Limits*

Future Titles Include:

Insulin Stability

Electrolytes

Sleep and Recovery (Vitality Series)

Strength and Aging (Vitality Series)

Part of a Broader Stability Framework

The Metabolic Stability Series is part of a larger stability ecosystem published by Perennial Publishing.

Related domains include:

Money Stability — structural durability in financial systems

Career Stability — positioning, load tolerance, and transition resilience

Mental Stability — cognitive steadiness under pressure and uncertainty

Emotional Stability — regulatory steadiness in environments that do not slow down

Foundations (Foundational Living Systems) — low-fragility life architecture

Each domain approaches stability as a structural problem: capacity versus load, buffering versus shock, governance versus drift. Together, they form a unified framework for long-horizon durability across the major systems that shape human trajectory.

About Perennial Publishing

Perennial Publishing produces practical systems guides for building stability in an unstable world.

Our books explore the foundations of long-term resilience across health, finance, career, and everyday decision-making. Each title focuses on clear frameworks, durable principles, and practical strategies that help readers build lasting stability rather than short-term fixes.

The Perennial catalog includes series on metabolic health, financial stability, career systems, and personal consistency. While each book stands on its own, they share a common goal: helping individuals create durable systems that support a stable and sustainable life.

For bulk or institutional orders, contact Perennial Publishing.

Made in United States
North Haven, CT
04 November 2026

21768648R00109